Safety & Risk – A 30-Day Mindset Transformation Guide

Daily Lessons in Evidence-Based Safety: Merging Human Leadership and System Discipline

Part of the *30-Day Mindset Transformation Series*

Other Books in the 30-Day Mindset Transformation Series

Safety & Risk – A 30-Day Mindset Transformation Guide

Daily Lessons in Evidence-Based Safety: Merging Human Leadership and System Discipline

Andy Page, Ph.D.

Published by EBR Technologies

ISBN: 979-8-9939275-0-3

Printed in the United States of America
Published by EBR Technologies

Disclaimer

This book is intended for educational and professional development purposes. The concepts, methods, and examples presented reflect the author's experience and interpretation of best practices within the fields of maintenance, reliability, and organizational culture.

While every effort has been made to ensure accuracy and clarity, the information provided is not a substitute for sound engineering judgment, professional advice, or site-specific analysis. Readers are encouraged to adapt these ideas to their own organizations responsibly, with appropriate technical validation and safety consideration.

Neither the author nor EBR Technologies, LLC assumes any responsibility for outcomes resulting from the application of the material in this book. All implementation decisions remain the sole responsibility of the reader and their organization.

Names and examples of companies, individuals, and situations have been used for illustrative purposes only and do not represent actual entities or events unless explicitly stated.

Table of Contents

Dedication

For the ones who stay awake a little longer after a near miss, replaying what could have gone differently.
For the supervisors who carry the weight of someone else's close call as if it were their own.
For the leaders who choose to care even when it's inconvenient — who know that every shortcut taken today becomes tomorrow's story.
For the people who refuse to treat safety as a number, a rulebook, or a program — because they understand it's a promise.

This book is for those who believe that safety is not managed — it's modeled.
That culture isn't built by compliance — it's built by conscience.
That the highest form of leadership is consistency — especially when no one is watching.

To everyone who treats every shift as an act of stewardship:
this book is for you.

Abstract

Safety doesn't fail by accident — it fails by drift.
This book reveals how to bring that drift under deliberate control. In thirty days of focused reflection and practice, The 30-Day Safety Mindset helps leaders close the gap between what they say they value and what their systems actually prove.

Each daily lesson explores one dimension of alignment — where belief meets behavior, where intention becomes inspection, and where leadership tone turns into system discipline. Grounded in psychology, neuroscience, and decades of field experience, this book connects the human side of safety with the structural side of control. It shows how communication, clarity, and consistency create the rhythm that reliability depends on.

The result is a practical framework for cultural calibration — a way to shrink "Delta S," the gap between intent and outcome, policy and practice, promise and proof. Through short readings, reflections, and daily EBR Principles, readers learn to model composure, reward transparency, and treat safety not as a compliance activity but as an alignment discipline.

This book isn't about control — it's about coherence. It teaches how to make safety visible through rhythm, language, and example, how to design systems that mirror belief, and how to build trust through predictable behavior.

In the end, The 30-Day Safety Mindset is more than a guide — it's a language for alignment.

Author's Note

After decades of working in safety and operations, I started noticing a pattern: the organizations that truly improved weren't the ones that wrote the longest policies or launched the biggest safety campaigns—they were the ones that changed how they thought.

I've seen leaders install new systems, update procedures, and retrain entire workforces—only to end up right where they started. Not because the programs failed, but because the Mindsets never shifted. The words were right, but the reactions were wrong. The slogans were new, but the decisions were old. The system couldn't change because the thinking that ran it hadn't changed.

That realization led me here. I wanted to capture the Mindsets that make safety feel natural, not forced—where leadership doesn't have to chase compliance because the culture itself has already chosen care.

Each page distills a single principle — something you can test in thought, apply in behavior, and see reflected in your results. The experiment is you — the way you observe, decide, and lead. Every entry follows the same rhythm: a moment to pause, a mindset to challenge, and a discipline to practice. Because alignment isn't built in a day — it's built through repetition.

Think of each day as a small calibration — a reset before the day's noise begins. That's where transformation really happens.

Mindsets lead to behaviors. Behaviors lead to results.
If we want safer outcomes, we must first learn to think in safer ways.

Safety doesn't begin in systems or procedures — it begins in people.
It begins in how we think.

— Andy Page, Ph.D.
Founder, EBR Technologies

Foreword

Merging Human Leadership and System Discipline

Every safe organization runs on two invisible engines: one human, one structural.
The human engine is made of leadership, attention, and belief — it's powered by what people notice, say, and model when no one's watching.
The structural engine is made of systems, procedures, and checks — it's powered by what the organization standardizes, measures, and sustains.

When the two move together, reliability feels natural.
When they drift apart, safety becomes fragile.

Most companies work hard on one and underestimate the other. They write procedures but neglect presence. They train people but fail to reinforce systems. They teach compliance but never cultivate conviction. And so, every few years, the same cycle repeats: a new campaign, a new initiative, a new slogan that quietly fades because it never found traction where culture actually lives — in the space between *human leadership and system discipline.*

That's the space this book is written for.

The **30-Day Safety Mindset** isn't a manual or a checklist. It's a mirror.
Each day explores one small intersection between the inner world of leadership and the outer world of systems — the moment where behavior meets structure. Because that's where safety truly happens: not in documents, but in decisions. Not in slogans, but in habits.

Safety is more than the absence of injury; it's the presence of alignment.
Alignment between what leaders say and what they reward.
Between what procedures declare and what people actually do.
Between how we manage risk and how we manage relationships.

Every lesson in this book aims to close that gap.
It's written to help leaders of every level — from the front line to the boardroom — see how their tone, timing, and tolerance shape the very system they rely on. Each page asks a simple question:

Are your systems protecting your people — or are your people protecting your systems?

In the end, safety maturity isn't measured by how many programs we launch, but by how little distance exists between intention and reality.
That distance has a name: ΔS — the Delta of Safety.
The smaller the delta, the stronger the culture.

This book is about shrinking that space — until what we *require* and what we *reinforce* are indistinguishable.
When belief and discipline move in rhythm, safety stops being a project and starts becoming a property of how we work, lead, and live.

That's the goal of Evidence-Based Safety:
to merge human leadership and system discipline — until the two are one and the same.

How to Use This 30-Day Guide

A Daily Practice in Merging Human Leadership and System Discipline

This book is not a checklist. It's a calibration tool — a way to realign what you think, what you feel, and what you do about safety.

Each day introduces one essential connection between *people* and *process*, reminding you that safety is not built from programs or posters but from rhythm — the rhythm between leadership behavior and operational discipline.

You don't need to read it in a rush.
Start each morning before the noise of the day begins.
Read one entry slowly. Sit with it. Reflect.
By evening, notice where it showed up — in your choices, your tone, your habits.

Each day follows a simple, repeatable structure:

- **The Moment** – A short, story-like entry that captures a real-world scene. Something you've likely lived or led through before — the kind of moment that quietly defines culture.
- **The Mindset** – The belief underneath the moment. It names the mental model that drives what people do when no one's watching. It turns reaction into reflection.
- **The Discipline** – The system behavior that gives that mindset structure. It's where the psychology of leadership meets the process control of safety.
- **Reflection** – A question to sit with. It's not about agreement — it's about awareness.
- **Commitment** – A single, declarative statement to anchor your intent for the day. Read it out loud if you can. Let it become a small ritual of accountability.
- **EBR Principle** – A grounding truth from *Evidence-Based Safety*: a reminder that every belief, every behavior, and every result leaves evidence.

This rhythm matters more than speed.
The lessons are short by design — long enough to provoke change, but brief enough to finish before the first meeting.

You'll discover that some pages speak directly to your leadership habits; others to the design of your systems. The goal isn't to favor one over the other — it's to unite them.
Because safety doesn't live in programs or posters — it lives in patterns.
And every time you read, reflect, and realign, you reduce drift between what your culture *says* and what it *shows*.

That is the daily work of reliability.
That is the discipline of Evidence-Based Safety.
That is how you merge human leadership and system discipline — one mindset at a time.

Note on Bonus Materials in the Back of the Book

At the end of this book, you'll find a collection of bonus tools designed to support the work you do here — frameworks, checklists, reflection guides, and reference materials you can use long after the 30 days are complete. They're not required to follow the daily rhythm, but they're there when you're ready to go deeper.

Why the Emphasis on Mindset?

Every improvement journey begins with a decision — and every decision begins with a mindset. Before you can change what people do, you have to change what they believe about what matters. That's why every book in this series starts with thinking, not tools. Procedures and policies don't stick if the people inside the system still see the world the same way.

Mindset is the hidden architecture of behavior — the lens through which we interpret data, make judgments, and justify choices. When the frame is wrong, the evidence doesn't matter. Leaders often install new methods on top of old Mindset, then wonder why the change collapses. Methods manage behavior. Mindset determines it.

If you want different results, you must start upstream — with how people think about their work, their role, and their responsibility.

Mindset as the Bridge Between Human and System

Reliable organizations understand that mindset is the bridge between human leadership and system discipline. Systems give structure; Mindset give meaning. One without the other drifts.

Checks, audits, and communication loops only work when people believe the discipline itself matters. When people see safety or reliability as compliance, they work to avoid blame. When they see it as stewardship, they work to protect value. The process may look identical — but the mindset behind it changes everything.

Belief Before Behavior

Human performance research shows that people act their way into consistency, but believe their way into commitment. Beliefs shape what we notice, how we interpret risk, and what we feel responsible to do.

Change efforts built only on process often trigger resistance. When they begin with mindset, they invite reflection instead of defensiveness. Mindset work slows us down long enough to see our thinking — and once we can see it, we can choose it.

The Role of Reflection

Each day in this series uses reflection because reflection turns belief into evidence. It's easy to agree with a principle; it's harder to see where our own behavior quietly violates it.

The daily structure — The Moment, The Mindset, The Discipline, Reflection, Commitment — surfaces that gap. The goal is not guilt; it's growth. Real change comes from small, repeated recalibrations — a rhythm of awareness.

Evidence-Based Thinking

Mindset work is also evidence work. It replaces assumption with observation and story with pattern. When you treat your own reactions as data, you become both scientist and subject.

Evidence-based leaders look for behavioral patterns and adjust beliefs to match reality. These pages don't give rules; they offer mirrors — invitations to see clarity, consistency, or drift in your own environment.

The Outcome

When Mindset change, everything aligns faster. Communication sharpens. Systems gain purpose. Results finally match intent.

Mindset isn't the warm-up to the work — it *is* the work. It's how you close the gap between what you say and what you show. When leaders change how they think, they don't just create new systems — they create new possibilities.

The EBR Safety Framework

Safety as the reduction of uncertainty, the management of risk, and the engineering of predictable outcomes.

1. Safety Begins with Awareness

Awareness is the mature starting point.
It replaces assumption with observation and routine with vigilance.
Without awareness, safety becomes luck.
With awareness, it becomes discipline — the foundation of predictable work.

2. Safety Is Not the Absence of Incidents — It's the Absence of Uncontrolled Risk

Incidents are outcomes; risk is the reality underneath.
The leader's responsibility is to recognize hazards long before they become events.
Safety exists only when exposure is identified, understood, and controlled.
Control stabilizes performance.

3. Clarity Before Exposure

Most teams act before they fully assess.
Mature safety thinkers pause long enough to define the task, the hazards, the boundaries, and the conditions.
Actions taken without clarity are gambles, not protections.
Clarity reduces uncertainty, anxiety, and harm.

4. Conditions Over Comfort

Comfort feels safe.
Conditions tell the truth.
High-reliability teams observe what *is*, not what they hope it to be.
When conditions drive decisions, safety becomes intentional instead of accidental.

5. Every Risk Lives in a System

No hazard stands alone; it is shaped by procedures, tools, incentives, congestion, habits, and constraints.
Controlling the symptom without addressing the system guarantees recurrence.
When the system improves, many risks disappear at once.

6. Speaking Up Is a Signal of Strength, Not Defiance

Voices create visibility.
When people raise concerns, they are revealing blind spots — context, hazards, pressures, or changes you could not see.
Treat every concern as data, not disruption.

7. The Easy Explanation Is Usually the Dangerous One

The mind prefers simple stories.
Safety requires resisting them.
Digging deeper reveals contributing factors, hidden triggers, and subtle patterns.
Surface explanations protect feelings.
Deep explanations protect people.

8. Good Questions Prevent Bad Outcomes

Rushed answers create confidence without certainty.
Good questions expose overlooked hazards, assumptions, and dependencies.
Every prevented incident begins with a question that slows the pace just enough to see clearly.
The quality of questions shapes the quality of protection.

9. Controls Must Survive Challenge

A control that fails under scrutiny will fail in the field.
Challenge is not criticism — it is quality assurance.
Controls validated through questioning, verification, and testing are the only ones worthy of trust.
Rigorous review is respect for human life.

10. Protections Must Match Hazards

A protection is effective only when it directly addresses the hazard.
PPE is not a substitute for engineering.
Procedures are not a substitute for design.
Alignment between hazard and control is what separates lasting protection from temporary relief.
Elegance is accuracy, not complexity.

11. Safety Is a Loop, Not a Line

Identify → assess → control → verify → learn.
The cycle repeats because conditions change, tasks vary, and risk evolves.
High-reliability teams do not assume safety — they confirm it.
Loops create resilience.

12. Transparency Accelerates Protection

Hidden risks grow.
Transparency exposes reality early, invites insight, and removes the darkness where hazards flourish.
When information flows freely, controls strengthen and response time shrinks.
Truth is the fastest form of protection.

13. Safety Creates Ownership

When people participate in identifying hazards, shaping controls, and verifying conditions, they take ownership of the outcome.
Ownership cannot be commanded.
It emerges from involvement, clarity, and trust.
Where safety is shared, commitment is deep.
Where it is imposed, compliance is the ceiling.

SECTION 1 – FOUNDATIONS OF CONTROL

Focus:

Mindset, accountability, and management commitment — the human root of every safety system.

Every safety system begins with belief.
Before the first procedure is written or the first audit is performed, someone has to decide that safety matters — not as a rule, but as a value. That decision is the seed of every mature safety culture, and it grows through mindset, accountability, and commitment.

Mindset is the foundation. It shapes how leaders interpret risk, react to incidents, and respond to people. A leader who sees safety as a compliance burden will look for checkmarks. A leader who sees safety as a moral and operational responsibility will look for patterns. The same event — a near miss, a policy violation, a shortcut — triggers completely different actions depending on the mindset behind it. That's why the first failure in most safety systems is not mechanical; it's mental.

Accountability turns mindset into motion. It's the bridge between belief and behavior — the visible proof that leadership means what it says. Accountability doesn't start in a meeting; it starts in the mirror. It's the willingness of leaders to ask, *"What did I teach through my reaction?"* before asking, *"Who made the mistake?"* True accountability flows downward only after it flows inward. When leaders hold themselves to the same standard they require of others, trust expands. People stop hiding problems because they no longer fear being blamed for them.

Management commitment gives accountability its strength. Commitment is not about signing a policy — it's about defending a principle. It means safety remains a priority when production is behind, when costs rise, when pressure builds. Visible commitment creates cultural gravity: what leadership emphasizes repeatedly becomes what people believe is real. Employees don't measure safety commitment by posters or programs; they measure it by consistency.

Every audit, procedure, and control ultimately depends on this human root system. Without leadership mindset, accountability, and commitment, safety structures become hollow — technically correct but emotionally irrelevant. But when these human elements align, safety systems come alive. Rules gain meaning. Metrics gain honesty. And people begin to treat safety not as an obligation to enforce, but as an expression of who they are.

Because in the end, safety isn't built on paperwork or policies. It's built on people — and people follow what their leaders truly believe.

Day 1 — The Moment of Control

The Moment

Every job begins with a decision point — a quiet, almost invisible moment where someone chooses between control and convenience.
A mechanic decides whether to isolate a line before tightening a fitting. An operator decides whether to bypass an interlock "just for a second." A supervisor decides whether to start a task before the permit is signed because "it's the same as yesterday."
Nothing catastrophic happens — not today. The system holds, production continues, and everyone breathes easier. But something deeper shifts: a new standard gets written into the culture. The message is subtle but powerful — *control is flexible when time is tight.*

That is how drift begins. Safety doesn't disappear in a single event; it erodes in fragments of permission. Each small compromise teaches the next one to feel normal. Control isn't lost all at once — it's traded away in pieces, disguised as efficiency.

The Mindset

Control is not the opposite of freedom; it's the proof of respect.
It's the understanding that risk doesn't care about our intentions — it only responds to our discipline. Mature leaders know that the moment you relax control, you redefine what "normal" means for everyone watching. People imitate what keeps them safe *and* what gets them praised.

That's why control must start in the mind before it shows up in the method. It's an attitude that says, *I will protect the standard even when it slows me down.* Leaders who model that patience build teams that mirror it. They turn caution into culture.

When control feels optional, chaos hides behind competence. When control feels honorable, safety becomes emotional — not procedural. The right mindset transforms discipline from something enforced into something chosen.

The Discipline

Control lives in behavior, not in binders. It's not a rulebook — it's rhythm. It's the pause before action, the verification before assumption, the checklist done even when no one's checking. It's the willingness to explain *why* each control exists, so that rules feel like protection, not punishment.

Disciplined leaders embed control into the flow of work. They create predictable routines — pre-task briefings, lock-outs, line breaks, hand-offs — that make safety visible. These aren't bureaucratic hurdles; they're the boundaries that make trust possible. The best systems don't rely on memory or heroics; they rely on rhythm.

When discipline becomes normal, compliance becomes effortless. The system starts running on integrity instead of enforcement.

Reflection

- Where in my own world have I confused "nothing went wrong" with "everything was under control"?

Commitment

- Today, I will treat every act of control — every check, pause, and verification — as a leadership statement.
- I will prove that discipline still defines this team, especially when pressure tempts convenience.

EBR Principle

Control is not what leaders impose on others — it's what people learn to impose on themselves.
Every pause for verification is an investment in predictability, and predictability is the foundation of trust.

Day 2 — Nothing Happens by Itself

The Moment

Walk through any safe facility, and it's tempting to believe safety just *exists*. Equipment hums smoothly, floors are clean, signage is clear, and procedures are posted where they should be. But look closer, and you'll see fingerprints everywhere — choices made, routines upheld, discipline maintained. What looks automatic is actually intentional.

Safety doesn't happen by itself. Neither does drift.
Every system is constantly moving in one direction or the other — toward order or toward entropy. The moment leaders stop inspecting, stop reinforcing, or stop caring, the system begins to unravel. And because drift is quiet, it often goes unnoticed until an event forces attention back to where it should have been all along.

In that moment, everyone remembers: what isn't actively managed will eventually be lost.

The Mindset

Mature leaders reject the illusion of automatic control. They understand that safety isn't self-sustaining — it's self-eroding. The default state of every process, every procedure, every habit is decay.
The mindset of control says, *If I want stability tomorrow, I must nurture it today.*

This mindset separates the reactive manager from the reliable leader. The reactive manager measures success by the absence of incidents. The reliable leader measures success by the presence of behaviors that prevent them.

Nothing maintains itself — not trust, not procedures, not culture. Systems drift because people get busy, familiar, or fatigued. A leader's role is to notice that drift early, to pull attention back to process before the process pulls away from safety.

The Discipline

The discipline of prevention begins with rhythm.
Control meetings start on time. Walkdowns happen daily, not occasionally.
Communication loops close visibly. Audits aren't surprises; they're expectations.

Disciplined leaders don't wait for incidents to reveal weakness — they go looking for it. They understand that "nothing happening" is not proof of stability; it's an invitation to verify.
Their teams learn to see silence as data, not comfort. If nothing is being reported, they ask *why*. If everything feels calm, they ask *what's drifting beneath the surface*.

The discipline of control is the discipline of curiosity.
It's the continuous act of proving that the system still works.

Reflection

- Where am I assuming things are under control simply because nothing has gone wrong lately?

Commitment

- Today, I will stop treating calm as confirmation. I will treat silence as a signal to look closer.

EBR Principle

Reliability — and safety — are not steady states; they are sustained states.
What is not reinforced will regress. What is not checked will drift.
Nothing happens by itself — including control.

Day 3 – What You Tolerate, You Teach

The Moment

There's a quiet instant after every unsafe act when a leader decides what happens next. Someone bypasses a guard, skips a checklist, or shrugs off PPE "just this once." The job keeps moving, the schedule stays intact—and a silent message is delivered: "This is fine." What's tolerated today becomes tomorrow's truth. Culture doesn't need a memo to change; it only needs permission disguised as silence.

Every standard decays at the speed of convenience. Teams don't rebel against safety rules—they simply take their cue from how seriously leadership defends them. When supervisors walk past small violations without a word, the system starts taking its shape from what's ignored instead of what's written. The next person sees it, copies it, and soon the "exception" becomes the expectation. Drift doesn't announce itself; it whispers through repetition.

Leadership isn't measured by how often you correct others—it's measured by what no longer needs correction because your standards are clear, consistent, and nonnegotiable. Every decision you make under pressure becomes part of your organization's unwritten code. The lesson is simple but relentless: your tolerance defines the true limit of your control.

The Mindset

The mature leader knows that authority doesn't come from the badge on the vest—it comes from the credibility of consistency. People trust what they can predict. They may not always agree with the standard, but if they know you'll hold it the same way every time, they'll respect it. That predictability becomes psychological safety in disguise; it tells the crew, "We work in a place where the rules don't change with the weather."

The mindset of ownership asks: what am I teaching right now through what I allow? Every decision communicates. Every silence instructs.

A culture of discipline begins when leaders stop mistaking leniency for loyalty. Correcting a teammate isn't betrayal; it's protection. The hardest lesson in leadership is choosing to be respected rather than being liked.

The Discipline

Define your red lines out loud. Don't assume people know where "unacceptable" begins—draw it visibly. If you must make an exception, document the why, set an expiration date, and communicate the plan to restore normal conditions. Build short, rhythmic conversations into your shift—pre-task briefings, stop-point verifications, end-of-day reviews. Every repetition teaches people that the standard isn't situational.

When you correct someone, explain the principle behind the rule. "This isn't about compliance—it's about keeping control." When you praise someone, make it equally specific: "You stopped the job when it didn't feel right—that's the kind of control that keeps us all safe." Every visible response becomes a data point in your culture's learning curve. Consistency is not rigidity; it's reliability applied to behavior. The goal isn't to police people—it's to protect patterns that protect people.

Reflection

- Where have I allowed convenience to define the standard? What lesson did my silence teach yesterday that I now have to unteach?

Commitment

- Today, I will make one expectation unmistakably clear and defend it with calm consistency—so no one confuses silence with approval.

EBR Principle

In safety, silence is a signal. Every unchallenged shortcut becomes tomorrow's system rule. What you tolerate, you teach.

Day 4 – Commitment in Action

The Moment

Commitment isn't declared in meetings; it's proven in motion. Everyone says safety matters—until it costs time. Everyone promises accountability—until the deadline bites. The real measure of commitment appears in the ordinary, unrehearsed moments when pressure and principle collide. The foreman who pauses the job instead of pushing through. The supervisor who takes the time to review a near miss before restarting the shift. The executive who walks the floor and asks not "How many hours?" but "How many hazards did we remove?" These are the quiet acts that separate slogans from systems.

Safety lives or dies in these micro-decisions. The organization will always believe what it sees. If leadership rushes, others will. If leadership reflects, others will. The team isn't waiting for another speech; they're waiting for a pattern they can trust.

The Mindset

Commitment in safety is not a campaign—it's a personal discipline that turns belief into rhythm. The mature mindset understands that repetition builds credibility. A leader who starts every day with a visible act of control—checking a permit, reviewing a critical lift, confirming barriers—teaches that standards are not seasonal.

This mindset views consistency as care. It recognizes that the crew draws emotional stability from predictability. When leaders respond the same way to the same risk every time, anxiety drops and focus rises. People begin to see safety not as inspection, but as assurance.

The opposite of commitment isn't defiance—it's drift. Drift happens when good intentions aren't supported by predictable behavior. A committed leader closes that gap by turning values into habits others can rely on.

The Discipline

Take commitment out of abstraction and put it on your calendar. Begin each shift with a deliberate act that demonstrates attention: review the day's critical task, confirm the permit conditions, verify the team understands what "done safely" looks like. End each shift with a short debrief: what changed, what we learned, what we'll improve tomorrow. Over time these bookends become the rhythm of reliability.

Document small wins and close small gaps. When someone raises a concern, respond within the same day—even if the full fix takes longer. "I heard you, and here's the next step" builds more trust than a polished memo delivered too late. Use visibility as a teaching tool: when standards are met, show gratitude publicly; when they're missed, address privately but transparently.

Most of all, keep your tone steady. A predictable response to error teaches people that truth is safe. Fluctuating reactions—calm one day, explosive the next—turn honesty into risk. Emotional consistency is operational consistency.

Reflection

- If someone shadowed me for a week, what would they conclude safety really means here? Would they see a value that guides every decision—or an obligation that waits its turn?

Commitment

- I will make my belief in safety visible every day through one consistent action—something so steady that people begin to expect it before I arrive.

EBR Principle

Culture doesn't change when leaders make promises; it changes when promises become patterns. Commitment is control made visible.

SECTION 2 – STRUCTURE & CLARITY

Focus:

Policies, principles, organization, and communication — how safety becomes a shared language.

Safety systems are built on belief, but belief alone can't hold structure. It needs form — the visible framework that turns intention into instruction and instruction into habit. That's what this section is about: giving safety a shape people can follow and a language they can share.

Structure is the skeleton of every dependable culture. Without it, even the best intentions collapse under pressure. It provides the boundaries where freedom operates safely — the lines that allow creativity without chaos. Clarity, meanwhile, is the nervous system. It carries the messages that coordinate motion and meaning. Together, structure and clarity give safety both posture and purpose.

Policies, principles, and organization charts aren't paperwork; they are architecture. They define how authority flows, how accountability is assigned, and how communication circulates. But those frameworks only live when people inhabit them. A structure on paper is a diagram. A structure in action is culture.

This section explores the disciplines that transform safety from a set of documents into a way of working. It begins with policies and principles — the written commitments that express what the organization stands for. Then it examines structure — how those principles are organized so everyone knows their role in protecting what matters. From there, it explores communication — the lifeblood of alignment, where leadership intent becomes shared understanding.

The goal isn't bureaucracy; it's coherence. The clearer the system, the less energy people waste interpreting it. When roles, expectations, and messages are unambiguous, employees spend less time guessing and more time executing. Confusion is one of safety's most dangerous hazards, and clarity is its simplest control.

The test of strong structure isn't how well it looks in an audit, but how calmly it functions in a storm. During a crisis, people revert to what's clear and trusted. That's why this section calls leaders to write, speak, and organize in ways that endure — where structure supports consistency, and clarity protects confidence.

When safety becomes a shared language, it stops being a program. It becomes a conversation — fluent, familiar, and alive..

Day 5 – Write It Down, Live It Out

The Moment

Every strong system begins with a sentence someone cared enough to write down. A policy isn't just paperwork—it's a declaration of value, a line drawn between what we'll accept and what we won't. Yet in many organizations, those words live in binders instead of behavior. The gap between what's written and what's witnessed becomes the first fracture in credibility. People see the posters, but they also see the exceptions. When the two don't match, the message dies quietly.

A safety policy is only as powerful as the evidence that supports it. The crew doesn't read the manual to decide what matters—they watch how leaders behave under pressure. Every time a supervisor overrides a rule to meet a deadline, a new "unwritten policy" takes shape. Before long, that version becomes the real one.

The Mindset

The mature mindset sees documentation not as bureaucracy, but as clarity. Writing down principles isn't about control—it's about consistency. A clear policy protects judgment; it gives people a common reference when emotions run high or decisions get rushed. It doesn't remove accountability—it defines it.

Policies are promises in print. They set expectations for both sides of the relationship: leadership promises to uphold them, and the workforce promises to follow them. When those promises are visible and reliable, trust grows. When they're vague or selectively enforced, cynicism spreads.

The mindset of "write it down, live it out" means this: every word you approve, you own. You can't delegate belief. If it's written in your system, it must be visible in your schedule, your spending, and your behavior.

The Discipline

Revisit every policy as if it were new. Ask: does this rule describe what we actually do, or what we wish we did? Rewrite until it matches reality—or change the reality until it matches the rule.

Then prove it. Take one principle each week and make it visible. If the policy says, "Stop work if unsafe conditions exist," show your team what that looks like in practice. Pause a job. Explain why. Reinforce that safety isn't a slogan—it's a system of supported choices.

Simplify the language. Replace technical clutter with words that teach. People can't follow what they can't remember. Clarity is kindness.

And when you see behavior that aligns with policy, acknowledge it out loud. Recognition is how policies stay alive. Every public reinforcement turns written intent into shared identity.

Reflection

- If an outsider followed us for a week, would they recognize our stated policies in our daily actions?
- Or would they discover that our real policies live in the hallway conversations instead of the handbook?

Commitment

- I will translate one written standard into visible action today—so people can see that policy is not paper; it's principle in motion.

EBR Principle

Documentation isn't bureaucracy—it's evidence of belief. What's written should always be what's real.

Day 6 – The Shape of Safety

The Moment

Every organization already has a safety structure—it's just not always the one drawn on paper. The real structure reveals itself in how people respond when something goes wrong. Who gets the first call? Who makes the decision to stop work? Who feels responsible when no one's watching? That living map of accountability and communication defines the *shape of safety*.

When structure is clear, safety flows naturally. Everyone knows their lane, their authority, and their role in protection. But when roles blur, confusion fills the gap. Responsibility scatters, and so does ownership. Tasks are done, but not connected. Risks are spotted, but not solved. The organization becomes reactive instead of resilient—not because people don't care, but because the structure doesn't care for them.

The Mindset

The mature mindset understands that safety is a team sport, not a department. It's not "their job" or "our job"—it's *the* job. Titles and charts are useful, but clarity is sacred. People need to know exactly where they fit and who depends on them. Structure isn't about control; it's about connection.

This mindset treats structure as an enabler of trust. When everyone knows who owns what, uncertainty disappears. The operator trusts the supervisor to act. The supervisor trusts engineering to respond. The system works because everyone knows how to hand safety forward. Confusion is the enemy of accountability; clarity is the foundation of control.

Leaders with this mindset constantly test for clarity: "Who owns this hazard? Who signs off on this permit? Who checks it last?" Each question strengthens the chain. The goal isn't hierarchy—it's harmony.

The Discipline

Strong structures don't happen by accident; they're engineered. Start by mapping your safety flow the same way you'd map a process. Identify every decision point: hazard identification, job planning, permits, audits, investigations. Then assign ownership by role—not name—so responsibility survives personnel changes.

Make accountability visible. If you lead a team, your people should know exactly what safety performance you're measured on. If you're part of a team, you should know how your work connects to the next link in the chain.

Hold cross-functional safety reviews. Bring maintenance, operations, and safety together not to defend, but to design. When everyone sees the full system, the blind spots shrink.

Finally, keep your structure alive. Review it after every major event or change. Ask: "Did this structure hold under pressure?" When people skip the process, it's not rebellion—it's feedback. The structure needs repair.

Reflection

- If our organization's chart disappeared tomorrow, would people still know who owns safety in each moment?
- Or have we mistaken boxes and titles for clarity and trust?

Commitment

- I will define one link in the chain today—clarifying who owns it, who supports it, and how others depend on it.

EBR Principle

Structure doesn't limit safety—it enables it. Confusion is chaos in disguise; clarity is the calm before excellence.

Day 7 – Message Over Medium

The Moment

Every safety message competes for attention in a crowded room of deadlines, alerts, and distractions. But communication isn't measured by how loudly it's said — it's measured by what's understood, remembered, and acted upon. The real test of a safety message isn't whether it's sent; it's whether it *sticks*.

Too often, organizations confuse information with communication. They issue bulletins, hold briefings, send emails — yet the message dissolves before it reaches the floor. The problem isn't technology or tone; it's translation. When people don't see the message connected to their moment, it becomes background noise. The more noise, the less meaning.

Communication that protects is always personal. It moves from instruction to intention — from "follow the rule" to "this is why it matters." Until a message makes emotional sense, it won't make behavioral change.

The Mindset

The mature mindset understands that safety language is culture's operating system. Every word either programs trust or introduces doubt. Communication isn't a task — it's a transfer of belief. When leaders speak with clarity, humility, and purpose, people don't just listen; they align.

This mindset sees communication as a shared responsibility, not a broadcast. A briefing is only half complete when it's delivered; it's finished when it's understood. Mature communicators check comprehension, not compliance. They ask, "What did you hear?" instead of "Did you hear me?"

They also understand that tone carries truth. Urgency without panic. Authority without arrogance. Empathy without excuse. When tone and intent match, communication feels safe — and safety becomes believable.

The Discipline

Start every communication with purpose. Before sending, saying, or posting anything, ask: "What do I need people to *understand* — not just know?" If the message doesn't change perception or behavior, it's not communication; it's noise.

Simplify relentlessly. Replace jargon with meaning. "Zero harm" isn't a slogan — it's a promise that must be explained through action. Speak in examples. Tell short, specific stories that anchor abstract principles in daily reality. People remember what they can picture.

Create feedback loops. Ask open questions: "What risk did we miss?" "What made this message unclear?" Communication that only flows downward will eventually stop flowing at all. When people see their voices shape decisions, they stop filtering truth.

Finally, model message discipline. Don't say what you can't support, and don't support what you won't say. Consistency is communication in motion.

Reflection

- Do my words create clarity or clutter? When I speak about safety, do people feel informed — or inspired?

Commitment

- I will make my next safety message so clear that it could travel without me — understood, not just heard.

EBR Principle

The medium delivers sound. The message delivers meaning. The leader delivers belief.

Day 8 – The Sound of Safety

The Moment

Every organization has a sound — a background hum that tells the truth about its culture. You can hear it in the way people start meetings, report issues, and talk about mistakes. The tone of those conversations says more about safety than any slogan ever will. When the sound is rushed, defensive, or silent, safety is being performed, not practiced. When it's calm, curious, and consistent, safety is being lived.

The sound of safety isn't volume; it's harmony. It's the moment when message, method, and meaning align — when what's said, how it's said, and why it's said all point in the same direction. That's when people stop following rules out of fear and start following them out of belief.

The Mindset

The mature mindset listens for patterns, not perfection. It knows that culture isn't just seen — it's *heard*. Every question asked, every story told, every word repeated either reinforces trust or erodes it. Leaders with this mindset don't just talk safety; they tune it. They understand that communication is an instrument — one that requires balance between structure and emotion.

Policies without passion sound hollow. Passion without structure sounds chaotic. But when both meet — when a clear system is delivered through a steady human tone — safety becomes music people want to play.

This mindset also values listening as much as speaking. The mature leader recognizes that the most powerful messages are the ones shaped by feedback. Listening doesn't weaken authority; it strengthens alignment. When people feel heard, they hear better.

The Discipline

Pause and listen. Walk the floor without an agenda. Hear how people talk about safety when they don't think they're being evaluated. Their tone will tell you everything: Do they sound confident or cautious? Curious or cynical? Hopeful or hesitant? The culture speaks long before the metrics do.

Then, audit your own sound. Review recent messages, meetings, and responses. Did your tone create calm or confusion? Did your reactions invite dialogue or defensiveness? Consistency of tone is consistency of leadership.

Finally, close the loop. After any communication, follow up visibly. "Here's what we heard." "Here's what we changed." "Here's what stays the same." Those phrases build credibility faster than any campaign.

Reflection

- If our organization had no posters, slogans, or scoreboards — only conversations — what would people believe about our safety culture?

Commitment

- I will listen today not for noise, but for patterns — the tone that tells the truth about our culture.

EBR Principle

Safety isn't silent — it speaks through tone, timing, and truth. The sound you make today becomes tomorrow's standard.

SECTION 3 – COMPETENCE & CONDITION

Focus:

Training, procedures, and mechanical integrity — transforming knowledge into reliability.

Every system, no matter how well designed, is only as dependable as the people who operate it and the conditions that support them. Competence and condition are the two pillars that turn safety from an aspiration into a stable reality. One shapes understanding; the other sustains trust.

Competence is more than skill — it's judgment under pressure. It's the ability to recognize when a rule applies and when a principle must guide. Trained behavior becomes instinct, and instinct becomes reliability. Competence is what allows people to perform safely when the unexpected happens. It bridges the gap between procedure and perception, between what's written and what's required in the moment.

Condition, in contrast, is the state of readiness. It's not just the physical condition of equipment or the workspace — it's the emotional and cognitive environment that surrounds work. Safe systems depend on more than tools that function; they depend on people who can think clearly, communicate openly, and trust the process around them. When either competence or condition deteriorates, risk accelerates.

Training and procedures are how organizations shape these two forces. Good training does more than transfer information — it transfers intent. It connects every task to purpose and principle, showing why precision matters as much as performance. Procedures, likewise, are not constraints; they are commitments — promises written down to ensure that good judgment has structure to stand on.

Mechanical integrity represents the visible side of this same equation. It's the assurance that what we depend on won't fail us without warning — that systems are designed, inspected, and maintained in a way that reflects the value of human life. When competence and condition intersect, integrity becomes predictable.

This section examines how to transform knowledge into confidence, and confidence into culture. It explores how learning deepens through repetition, reflection, and accountability — and how consistent care sustains trust in both people and systems.

Reliability, in this sense, isn't mechanical — it's moral. It's the outcome of disciplined preparation and visible respect for what can go wrong. Competence provides the skill; condition provides the stability. Together, they define the human texture of every safe operation.

Day 9 – Discipline Creates Calm

The Moment

Every safe system begins the same way — with structure that feels inconvenient at first and essential later. Procedures are not rules to restrict thought; they are boundaries that protect it. They turn uncertainty into confidence by defining what *right* looks like before the chaos begins.

The calm you see in mature organizations doesn't come from luck. It comes from discipline — the willingness to repeat the right things the right way, even when no one is watching. Discipline is not control for control's sake; it's the quiet order that keeps people free to focus on higher decisions. It allows teams to move with precision instead of hesitation, to act with clarity instead of guessing.

The Mindset

The disciplined mindset views standards as safeguards, not burdens. It understands that freedom without form becomes fatigue — that when everyone invents their own version of "how," the result is noise, not progress. True professionals take pride in procedure because it reflects a deeper value: consistency.

Following a standard doesn't mean losing creativity. It means earning trust. Once the fundamentals are reliable, the mind is free to innovate safely. In this way, discipline and creativity are not opposites — they're partners. The foundation of great performance is the confidence that the basics will hold.

Mature organizations see procedures as promises. They say, "We've thought this through. We've learned from experience. We care enough to write it down." Those documents don't kill initiative; they preserve wisdom. Each step was written in the language of lessons learned — a record of what went wrong, so it doesn't have to again.

The Discipline

Every procedure should teach clarity, not compliance. When people understand the *why*, the *how* becomes meaningful. Leaders must ensure that standards are living documents — visible, reviewed, and explained through conversation, not merely enforced through audits.

The test of discipline isn't how tightly people are controlled but how confidently they perform under pressure. When procedures are written well and practiced often, teams instinctively fall back on them when things go sideways. That instinct — calm, measured, precise — is the real proof of discipline.

Reflection

- Are our procedures teaching clarity or just compliance?
- Do they calm people under pressure, or do they add confusion when speed matters most?

Commitment

- I will follow the standards not because I'm told to, but because I understand what they protect.

EBR Principle

True discipline doesn't confine people — it steadies them. Calm is the reward for consistency.

Day 10 – Teaching What Matters

The Moment

Training is not about filling time — it's about transferring belief. Every skill we teach carries a message about what we value. When we train people only on tasks, we teach that performance is enough. When we also train them on purpose, we teach that judgment matters just as much as execution.

The measure of great training isn't how much information was presented — it's how much understanding was absorbed. Too often, training becomes an event to be checked off instead of an experience that shapes behavior. But every time someone steps into a task without confidence or context, the system pays the price.

The Mindset

Teaching what matters begins with knowing what matters. Mature safety cultures don't start with regulations; they start with reflection. They ask: what are the behaviors, attitudes, and decisions that keep people safest? Then they train those, deliberately and repeatedly.

Good training connects the *what*, the *how*, and the *why*. The *what* gives knowledge. The *how* builds competence. The *why* builds conviction. When all three align, people don't just remember the steps — they internalize the reason. That's when culture begins to sustain itself.

Leaders play a quiet but powerful role here. Every correction, explanation, or story becomes a lesson. Every time a supervisor takes the extra two minutes to explain *why a shortcut matters*, that's training. Every time someone models calm under pressure, that's training. Culture learns through observation long before it learns through policy.

The Discipline

Effective training is never an accident — it's a designed rhythm. It doesn't fade once orientation ends. It's reinforced in pre-shift huddles, toolbox talks, mentoring moments, and debriefs after near misses. The most reliable systems are those where training doesn't feel like a classroom; it feels like conversation.

Development goes beyond skill to readiness. A well-trained person knows how to perform; a well-developed person knows how to respond. That difference is what keeps teams composed when conditions change. When people are trained for context, not just for compliance, they adapt safely instead of improvising dangerously.

Reflection

- Are we training for knowledge or for judgment?
- Do our lessons end when the slides do, or do they live in the way our people think and act every day?

Commitment

- I will teach not just what to do, but why it matters — because safety learned without context is safety soon forgotten.

EBR Principle

Training transfers more than skill; it transfers belief. Every lesson taught is a value declared.

Day 11 – The Integrity of Things

The Moment

Every system eventually reveals what it's made of. Materials, processes, and people all share one truth: integrity isn't proven in calm; it's proven under stress. The structures we build — physical or cultural — must be designed to hold when pressure rises. That's what mechanical integrity really means: confidence that what we depend on will perform when we need it most.

In safety, integrity begins long before inspection. It begins with intention — the decision to design for reliability, to maintain for continuity, and to operate with care. The quiet work of preserving condition is not glamorous, but it's the heartbeat of trust. When people see equipment that's cared for, standards that are upheld, and systems that don't cut corners, they don't just believe in the machine — they believe in the culture behind it.

The Mindset

Integrity is both technical and moral. It's not just about bolts, valves, or welds; it's about the invisible forces of attention and accountability. Machines don't fail because they forget — they fail because people do. When the culture tolerates neglect, every unchecked gauge and every postponed inspection becomes a vote for risk.

A mature organization treats every asset — mechanical or human — as a reflection of its values. They don't wait for failure to remind them what discipline prevents. Integrity is the art of prevention disguised as routine. It's the daily repetition of care that ensures stability tomorrow.

When people see leaders take interest in condition — walking the floor, asking questions, respecting the process — it elevates the work. It turns maintenance into stewardship. It shows that the organization doesn't just talk safety; it protects it through deliberate attention.

The Discipline

Mechanical integrity isn't just about doing inspections — it's about doing them *with meaning*. Each inspection, calibration, or test is a reaffirmation of control. Integrity lives in those quiet verifications that prove trust is deserved. The disciplined leader doesn't see inspection as distrust; they see it as commitment — a way to keep promises visible.

Systems decay when vigilance becomes optional. Once "good enough" replaces "verified," drift begins. The antidote isn't fear — it's pride. The team that treats inspection as ownership instead of obligation will never fall far from reliability.

Reflection

- Are we maintaining condition with purpose, or just performing the motions of control?

Commitment

- I will protect the integrity of the system by taking pride in the details others overlook.

EBR Principle

Integrity isn't just strength — it's attention made visible. What you inspect faithfully, you preserve.

Day 12 – Review: Capability and Care

The Moment

Competence and care form the twin pillars of safety. One without the other creates imbalance. Skill without care becomes mechanical; care without skill becomes fragile. The safest cultures are those where ability and attention grow together — where knowing how is inseparable from remembering why.

Across the last three days, we've looked at the discipline of procedure, the shaping power of training, and the quiet assurance of integrity. Each theme points to a deeper truth: safety is not a reaction to risk; it's a reflection of readiness. Readiness is what happens when people and systems are both prepared — technically, mentally, and emotionally — to meet the demands of reality.

When we teach with purpose, we build judgment. When we follow procedures with respect, we build trust. When we maintain condition with pride, we build credibility. Those three actions — teach, follow, maintain — are how safety earns its voice. They are the grammar of discipline.

The Mindset

Competence is more than qualification. It's confidence anchored in evidence. It's the feeling of control that comes from knowing you've done the right things, the right way, for the right reasons. When people are trained well, supported well, and equipped well, they stop guessing and start deciding. And that shift — from reaction to intention — is where maturity begins.

But competence alone doesn't guarantee safety. Without care, it drifts into compliance. People perform steps without meaning, and meaning is what gives structure its strength. Care is what keeps procedures alive. It's what turns inspections into stewardship and repetition into rhythm. The best leaders never let competence overshadow compassion. They remind their teams that skill is a gift — and that protecting others through that skill is an act of pride.

The Discipline

Leaders reinforce capability not by testing people, but by trusting them with purpose. They don't just assign tasks; they invest context. They don't measure hours; they measure understanding. Every meeting, briefing, or inspection becomes a moment to prove that control doesn't limit freedom — it sustains it.

True capability is demonstrated when no one is watching, and true care is revealed when no one reminds you. When those two align, the system becomes self-correcting. People see what's right, not just what's required.

Reflection

- Do our systems build both competence and care?
- Are we teaching people to understand, or simply to comply?
- What would happen if we treated every task as a chance to demonstrate both mastery and meaning?

Commitment

- I will practice safety as both precision and compassion — ensuring that what I know serves what I care about.

EBR Principle

Skill creates confidence; care creates connection. When the two are synchronized, safety becomes second nature.

SECTION 4 – OBSERVATION & LEARNING

Focus:

Audits, investigation, and motivation — using evidence, not emotion, to drive improvement.

Observation is the beginning of understanding. In every safety system, what we choose to see — and how we choose to respond — determines whether we improve or repeat. This section explores the discipline of noticing, the courage to inquire, and the humility to learn. These are the quiet traits that separate reactive organizations from reflective ones.

Observation isn't surveillance; it's stewardship. It's how leaders stay connected to truth. A walk-through, an audit, a conversation at the job site — these moments aren't about catching mistakes; they're about catching drift before it becomes danger. They remind people that attention is care made visible. When leaders observe, they signal what matters. When they ask questions instead of issuing orders, they teach curiosity instead of compliance.

Learning begins where blame ends. Accidents and near misses often expose not bad people, but broken systems — missing clarity, weak reinforcement, or flawed assumptions. The mature leader resists the temptation to assign fault and instead seeks cause. They treat each deviation as data, each error as a clue. In this way, investigation becomes less about punishment and more about pattern recognition — the evidence that guides prevention.

Motivation ties it all together. People don't improve because they're afraid; they improve because they're inspired. Fear silences truth, but motivation draws it out. When individuals see learning as safety's highest form of strength, they begin to speak openly, share lessons, and engage in problem-solving. The organization grows not by enforcing perfection, but by rewarding honesty.

In practice, this section moves from the mindset of "finding fault" to "finding cause." It challenges the reader to see audits as feedback loops, not scorecards; to see investigations as acts of care, not correction; and to view motivation as a tool for alignment, not control.

Because the safest systems are not the ones that never fail — they're the ones that never stop learning. Observation gives us truth. Investigation gives us understanding. Motivation gives us momentum. Together, they turn experience into intelligence, and intelligence into prevention.

Leadership Truth: Every observation is a choice between judgment and learning. The culture that chooses learning will always emerge stronger than the one that chooses blame.

Day 13 – Learning Before Blaming

The Moment

Something goes wrong. A near miss, a deviation, a close call that could have been worse. The first question reveals the culture: *Who did it?* or *What happened?* That single word — *who* or *what* — decides whether an organization learns or just looks for someone to blame.

In too many systems, the reflex is accountability through accusation. We confuse fault with cause, emotion with evidence. But every incident tells a story, and that story deserves to be understood, not edited to fit our assumptions. When we look for fault, we stop too soon. When we look for cause, we start to grow.

The most mature safety cultures understand that people rarely fail on purpose. They fail because the system allowed them to. The procedure wasn't clear. The signal wasn't visible. The standard wasn't reinforced. When a leader sees that — when they realize the mistake is often a symptom of a missing safeguard — everything changes. The investigation becomes a process of learning, not judging.

The Mindset

Curiosity is the highest form of leadership discipline. It takes strength to ask questions when blame would be faster. It takes patience to listen to a story you think you already understand. But the moment you stop listening, you stop leading.

Blame teaches silence. Learning teaches truth. People only speak when they feel safe, and safety doesn't mean the absence of consequence — it means the presence of fairness. When leaders respond with perspective instead of punishment, they create psychological safety — the invisible condition that allows honesty to thrive.

Every organization teaches its people how to tell the truth. If the truth is punished, it will hide. If the truth is valued, it will multiply. The choice belongs to leadership.

The Discipline

Learning cultures treat every event as a data point, not a disgrace. They document, discuss, and decide together. The investigation becomes a mirror — one that reflects not just what happened, but what leadership allowed to happen.

The disciplined leader uses evidence as a teacher. They ask: *What signals were ignored? What conditions made this likely? What can we change so it never happens again?* These are not soft questions — they're structural ones. And they build maturity faster than any reprimand ever could.

Reflection

- Are we investigating to understand or to defend?
- Do we treat mistakes as human failures or system failures?
- What message does our response send about what we truly value?

Commitment

- I will respond to failure with curiosity, not accusation — because learning, not blaming, is what keeps people safe.

EBR Principle

Blame ends the story. Learning continues it. The system that seeks understanding will always find strength.

Day 14 – The Power of Watching

The Moment

Observation is leadership in motion. It's not about catching people off guard — it's about staying close enough to see reality as it truly is. When a leader walks the floor, visits a job site, or joins a pre-task talk, they're doing more than inspecting work; they're reinforcing worth. They're proving that safety isn't something delegated — it's something demonstrated.

Yet the word *audit* often creates anxiety. People brace for judgment, not conversation. That's because too many audits focus on fault-finding instead of fact-finding. The power of watching isn't in inspection — it's in interpretation. Great leaders don't walk around with a clipboard looking for what's wrong; they walk around with curiosity, looking for what's real.

Every system tells the truth to those who take the time to observe it. The noise, the shortcuts, the quiet workarounds — all reveal where friction lives. When a leader sees patterns without prejudice, they uncover not just compliance gaps but design gaps. And that kind of seeing — patient, humble, and evidence-based — becomes one of the most powerful safety tools of all.

The Mindset

Observation is communication. What you notice, you normalize. When you walk past a hazard without reaction, you've taught tolerance. When you pause to recognize a safe act, you've taught importance. People watch leaders more than they listen to them. The tone of your attention tells the story of your priorities.

Mature organizations turn observation into a shared language. They teach everyone — from operators to executives — how to watch with purpose. That means noticing patterns, not just incidents. It means asking "why" with interest, not accusation. And it means treating every observation as an opportunity to strengthen connection, not control.

Observation done right builds mutual respect. It tells employees, *you matter enough to be seen.* It tells leaders, *you're accountable for what happens when you look away.*

The Discipline

Effective audits are consistent, transparent, and fair. They measure process health, not personal worth. They include follow-up and feedback, so people see that what was observed was also understood. Without that closure, audits feel like theater. With it, they become transformation.

The disciplined leader doesn't audit to prove compliance — they audit to preserve credibility. They understand that the act of watching, when done with empathy and rigor, turns oversight into insight.

Reflection

- When I observe work, am I searching for mistakes or understanding the system?
- Does my presence build trust — or tension?
- What do my eyes teach when I look around?

Commitment

- I will observe to understand, not to judge — because every observation is a chance to build connection, not control.

EBR Principle

Observation without empathy is inspection. Observation with empathy is leadership.

Day 15 – Motivation Without Fear

The Moment

Fear is an effective motivator — but only once. It can spark action in the short term, but it never sustains belief. When people comply out of fear, they do only enough to stay safe from punishment, not to stay safe in reality. The system runs on anxiety instead of ownership, and the moment the fear fades, so does performance.

True motivation grows from meaning, not management. It begins when people see how their work protects others, not just themselves. It deepens when leadership connects safety to pride, not paperwork. The most successful safety cultures don't ask for obedience — they invite contribution. They turn compliance into commitment by giving people a reason to care.

The Mindset

Motivation without fear is built on trust and transparency. When people understand *why* rules exist, they stop seeing them as restrictions and start seeing them as protections. They begin to interpret procedures as the organization's way of keeping promises — a mutual agreement that everyone goes home the same way they arrived.

Leaders who motivate through meaning don't need slogans or incentives. They build belief through example. A sincere thank-you after a job done safely can reshape morale more than a bonus ever could. People want to know they're seen, that their effort matters, that their vigilance is noticed. Recognition is fuel — but it must be authentic. Empty praise erodes faster than fear ever could.

Motivation is also contagious. Enthusiasm, care, and discipline spread just like frustration, apathy, and cynicism. The tone of a single leader can lift an entire shift or drag it down. That's why mature safety systems invest in emotional consistency as much as technical control. Calm breeds confidence. Panic breeds mistakes.

The Discipline

Motivation isn't a mood; it's a management practice. It requires regular reinforcement, open dialogue, and balanced accountability. Great leaders measure not just what people *do*, but what they *believe*. They ask questions that reveal ownership: "What's the most important safety choice you made today?" "What would you improve if you could?"

Motivation thrives where feedback flows both ways. When employees see their ideas adopted, their engagement becomes self-sustaining. They stop waiting for direction and start shaping direction. That's when safety shifts from a program to a principle.

Reflection

- What emotions power our safety culture — fear or pride?
- Are we inspiring action through belief or enforcing it through pressure?

Commitment

- I will lead by inspiration, not intimidation — because motivation that begins with meaning lasts longer than motivation that begins with fear.

EBR Principle

Fear enforces silence. Meaning invites ownership. The safest cultures are built on belief, not threat.

Day 16 – Review: The Evidence of Improvement

The Moment

Progress isn't proven by what people promise — it's proven by what they practice. The true measure of a safety culture isn't how few incidents it reports, but how much learning it produces. Mature organizations know that safety isn't the absence of accidents; it's the presence of insight. The goal isn't perfection — it's improvement made visible.

Every observation, investigation, and audit is a mirror. When we look into those reflections with honesty, we begin to see the system as it really operates. Patterns appear. Habits surface. Weak signals become visible before they turn into strong consequences. And in those moments, leadership's job is not to defend the system — it's to understand it.

The Mindset

Improvement begins where blame ends. When the focus shifts from "Who did it?" to "What does it teach us?", everything changes. Fear shrinks. Candor grows. People stop hiding problems and start helping solve them.

A culture that seeks evidence instead of excuses becomes self-correcting. It uses every event — even the painful ones — as data for better design. It replaces emotional reactions with rational reflection. And over time, it learns that accountability isn't punishment; it's ownership in practice.

The evidence of improvement isn't found in metrics alone. It's seen in behaviors: the supervisor who listens before judging, the operator who reports a near miss without fear, the manager who thanks someone for catching what could have gone wrong. These are the quiet signals of maturity — the daily proof that safety has become part of how people think, not just what they do.

The Discipline

Improvement without documentation is only memory. That's why evidence matters. Write it down. Track it. Share it. When lessons learned are captured, shared, and revisited, the organization starts to see itself clearly. Transparency doesn't weaken credibility; it builds it.

The disciplined leader treats learning as a process, not an event. Each observation feeds the next, each audit refines the next standard, and each investigation strengthens the next plan. Over time, improvement stops being something leadership demands — it becomes something the system expects.

Reflection

- Do our investigations end with blame or begin with learning?
- Can we point to visible evidence of progress — not just in reports, but in behavior?
- How do we know our system is improving, not just enduring?

Commitment

- I will make improvement visible — through honesty, documentation, and dialogue. The evidence of progress is not the absence of error, but the presence of learning.

EBR Principle

Improvement isn't declared — it's documented. The culture that learns out loud grows stronger with every truth it faces.

SECTION 5 – CHANGE & ADAPTABILITY

Focus:

Management of Change, quality assurance, and resilience — how cultures evolve without losing control.

Change has always been the most dangerous moment in any system. Most failures don't begin with recklessness; they begin with improvisation — a well-intentioned "fix" made by someone who thought they were helping. A pipe rerouted to save time. A control bypassed to keep production moving. A shortcut that worked once and quietly became normal.

That's why **Management of Change (MOC)** exists. It's the disciplined review process that separates ingenuity from risk. MOC isn't bureaucracy — it's protection. It ensures that every proposed modification is examined by the right people with the right technical competence before it touches the system. It asks the hard questions that enthusiasm forgets: What will this affect? What could this compromise? What must stay unchanged for the process to remain safe?

In a mature organization, MOC acts as a cultural speed governor — slowing down impulsive action just long enough for understanding to catch up. It keeps curiosity from outrunning competence. It ensures that what "seems like a good idea at the time" never becomes the root cause of the next investigation.

Quality assurance works in parallel. It confirms that every component, every material, and every configuration meets the standards that make safety predictable. Together, MOC and quality assurance form a double lock — one guarding the decision, the other guarding the design. Their combined purpose is simple: to keep the system coherent, no matter how much it evolves.

But structure alone isn't enough. A binder full of forms doesn't prevent accidents — disciplined participation does. MOC depends on people who respect process more than convenience, who recognize that pausing for verification is an act of stewardship, not delay. When operators, engineers, and leaders all treat change as something that must be *proven safe* before it is *made real*, culture reaches its highest form of maturity.

Adaptability, then, isn't about accepting change blindly — it's about engineering it wisely. A culture that treats MOC as sacred preserves both safety and innovation. Because the goal is not to stop change; it's to ensure that every change honors the integrity of what already works.

Leadership Truth: Change handled carelessly creates risk. Change handled wisely creates resilience.

Day 17 – Change Is a Test of Culture

The Moment

Change has a way of revealing what a culture actually believes. When a new idea is proposed — a shortcut, a design tweak, a control adjustment to "keep things moving" — the organization faces its most honest test. The question is no longer, *Do we care about safety?* but *How do we prove it?*

The moment between idea and implementation exposes whether safety is a system or just a slogan. Some cultures rush forward, convinced that good intentions equal good outcomes. Others pause, review, and confirm — not because they lack confidence, but because they understand that certainty must be earned. That pause is the difference between control and chaos.

Every process change, no matter how minor, is a pressure test of discipline. It asks whether leaders trust procedure more than persuasion, and whether employees feel supported in saying, "Let's check first." When the pressure rises, culture speaks louder than policy. What you tolerate in those moments — who gets to decide, how that decision is reviewed, and how it's communicated — determines not just safety performance, but trust itself.

The Mindset

Management of Change (MOC) is not red tape; it's engineered humility. It recognizes that even experienced people can't always see the second- and third-order effects of their decisions. MOC creates structured friction — a deliberate slowing-down of action so that reflection can catch up. It turns enthusiasm into analysis and converts motion into understanding.

This mindset values process over personality. It ensures that qualified reviewers assess every modification for technical soundness, operational impact, and human consequence before approval. It is the habit of asking, *What else will this touch?* and *Who needs to know before we act?*
Mature leaders model this by submitting their own ideas to review. They demonstrate that leadership isn't exemption from the process; it's devotion to it.

In doing so, they replace the myth of urgency with the discipline of assurance. The result isn't slower work — it's safer work that stays done.

The Discipline

Effective MOC is not paperwork; it's proof. Each review connects competence to accountability, preventing "good ideas" from becoming bad outcomes. When people see their leaders follow the same rules they enforce, credibility deepens.

Reflection

- Does our culture value the pause before change, or reward those who push past it?
- Do we treat MOC as protection or bureaucracy?
- What does our process say about what we truly believe?

Commitment

- I will never rush a change faster than it can be reviewed.
- I will prove that discipline is an act of care, not control.

EBR Principle

Evidence-Based Safety treats every modification as an experiment requiring validation. Verification is not delay — it is respect for the system and everyone who depends on it.

Day 18 – Designing for Confidence

The Moment

Confidence is not a feeling — it's a function of design. When systems fail, it's often because the organization believed everything was fine without proving that it was. Quality assurance exists to reverse that assumption. It turns hope into evidence. Every inspection, checklist, and verification is not bureaucracy; it's the visible expression of care.

In a mature culture, quality isn't an afterthought performed once a system is built — it's woven into how things are designed, fabricated, and approved. It's the principle that says, *Safety doesn't begin at startup; it begins at design.* The small decisions — the torque on a bolt, the calibration of a sensor, the wording of a procedure — carry consequences far larger than their size.

The moment of quality is the moment of truth. It asks: Have we verified what we think we know? Have we confirmed that what's "good enough" is truly safe enough? In that moment, discipline transforms intention into integrity.

The Mindset

Quality assurance is the conscience of the system. It refuses to assume that doing it once means it's right forever. It demands visible proof that what was promised is what was delivered. This mindset doesn't treat inspection as mistrust — it treats it as respect for complexity. It acknowledges that even the best teams miss things, and that catching them early is an act of protection, not punishment.

The mature leader never delegates quality as a nuisance or a separate department's burden. They own it as part of leadership's signature. They know that credibility lives or dies in the gap between "as designed" and "as built." When leaders model that attention, others follow suit. The conversation changes from "Who signed off?" to "How do we know?" — a shift from authority to evidence.

58

Quality assurance, done well, does not slow progress; it accelerates trust. It allows teams to move faster tomorrow because they are certain about today. It's how confidence becomes a collective property rather than an individual emotion.

The Discipline

Quality systems only work when their findings are respected. Every deviation logged, every test result questioned, and every failed inspection is data — not defeat. Discipline means responding with curiosity, not defensiveness. Over time, this builds a system that learns faster than it breaks.

Reflection

- Do we verify our confidence, or just assume it?
- Does our culture reward people who find problems or people who hide them?
- Are we as proud of our inspections as we are of our production?

Commitment

- I will make quality visible by insisting on proof, not promises.
- I will treat every verification as an act of respect for the people who depend on it.

EBR Principle

Evidence-Based Safety treats quality as confirmation, not compliance. Assurance is the discipline that turns good work into guaranteed safety — the bridge between belief and proof.

Day 19 – Seeing the System Shift

The Moment

Every system changes — the question is whether it does so by design or by drift. In the physical world, this might mean a modification to a facility, a new piece of equipment, or an update to a control system. In the organizational world, it's a policy rewrite, a new process, or a restructured workflow. Each of these changes seems small in isolation, but together, they alter how energy, information, and risk flow through the system.

The moment of change is not when the modification occurs; it's when someone decides it's safe enough to proceed. That's the crossroads where culture and engineering meet. Do we ask, *What could this affect?* or do we assume it's minor because it feels familiar? Most serious incidents begin not with the change itself, but with the failure to see its reach. The connections between systems — mechanical, procedural, or human — are almost always deeper than they appear. Maturity begins when organizations learn to see those connections before they feel them.

The Mindset

True safety culture sees systems as living organisms. Every change is like surgery — it requires preoperative planning, competent review, and post-change monitoring to ensure health has improved, not declined. This mindset values interdependence. It understands that a facility isn't a collection of parts; it's a web of relationships.

The discipline of **Management of Change (MOC)** for facilities and technology protects that web. It ensures that modifications, however well-intentioned, are evaluated by people with the right technical expertise. It demands that changes are recorded, reviewed, approved, and communicated before implementation. This isn't bureaucracy — it's collective memory.

Without it, organizations lose their ability to trace cause and effect. Technology adds another layer of risk. A simple software update can alter logic, timing, or data visibility in ways that the human eye can't detect. MOC ensures those invisible shifts are examined under the same rigor as physical changes.

The most dangerous phrase in safety is still, "It should be fine." MOC exists to prove that it actually is.

The Discipline

Strong MOC follows a rhythm: propose, review, approve, document, verify. Each stage reinforces clarity and accountability. When leaders refuse to bypass this rhythm — even under pressure — they teach everyone that stability matters more than speed. Every documented review becomes a safeguard against amnesia, ensuring the next generation inherits both the system and the lessons that shaped it.

Reflection

- Do we treat change as a design event or an interruption?
- How do we ensure every modification strengthens the system instead of surprising it?
- Where in our organization does "It's probably fine" still slip through?

Commitment

- I will see every change as a system event, not a local fix.
- I will protect the pause that ensures safety before speed.

EBR Principle
Evidence-Based Safety treats modification as a hypothesis that must be tested. Each verified review is a statement of integrity — proving that progress never comes at the cost of control.

Day 20 – Review: The Safe Path Forward

The Moment

Change always leaves evidence — not just in the form of new equipment, revised drawings, or updated software, but in the way people respond afterward. When a system change is well managed, confidence grows. When it's not, confusion lingers. The true test of culture is how an organization behaves once the modification is complete: Do we verify that the new system works as intended, or do we move on and assume it does?

The safe path forward begins with validation. Every change, no matter how well executed, introduces new interfaces — between machines, processes, or people.

These interfaces are where risk hides. A valve replaced, a setting adjusted, a sensor recalibrated — each carries the potential to create unanticipated interactions. Wise leaders know that post-change confirmation is not distrust; it's discipline. They treat re-startup not as a formality but as the moment the system earns back its reliability.

The Mindset

Safety cultures mature when they realize that "done" is not the same as "safe." The Management of Change process isn't complete until the organization has confirmed that outcomes match intent. This mindset sees follow-through as the heartbeat of professionalism. It replaces assumption with assurance and turns checklists into commitments.

Leaders play a critical role here. When they personally participate in post-change walkdowns, ask verifying questions, and require documented sign-offs, they demonstrate that thoroughness is not a burden — it's a value. Teams notice. Over time, those small demonstrations teach everyone that diligence is not optional; it's identity.

A culture that reviews its changes honestly is a culture that never stops learning. Instead of asking, "Did we finish it?" they ask, "What did this teach us?" That shift — from closure to curiosity — transforms MOC from an administrative step into a learning system. It's how small corrections prevent large catastrophes.

The Discipline

The discipline of follow-up gives change a memory. Each review, verification, and validation closes the loop of responsibility. The disciplined leader insists that evidence of success be as visible as the work that preceded it. That evidence — documented tests, updated drawings, sign-offs, and feedback — becomes the organization's proof of care.

Reflection

- After change, do we confirm or assume?
- What recent modification could benefit from a deeper post-review?
- Are our systems of assurance as strong as our systems of action?

Commitment

- I will make post-change validation a visible habit, not a hidden step.
- I will measure completion not by when the work stops, but when confidence starts.

EBR Principle

Evidence-Based Safety defines progress as verified stability. The path forward is not paved by speed, but by certainty — because confidence untested is only hope disguised as proof.

SECTION 6 – RISK & READINESS

Focus:

Risk analysis, emergency preparedness, and proactive safety leadership — closing the loop between anticipation and action.

Safety is not tested by calm days; it's revealed in moments of pressure. The measure of a mature organization isn't how well it prevents risk from appearing, but how well it anticipates and responds when it inevitably does. Every reliable system must live with uncertainty — and readiness is how it turns uncertainty into confidence.

Risk and readiness exist in partnership. Risk analysis is the art of prediction; readiness is the discipline of preparation. Together, they form the feedback loop that keeps a safety culture from becoming complacent. When people stop asking *"What could go wrong?"* or *"Are we ready if it does?"*, the organization drifts into fragility. But when those questions are embedded in daily routines — in planning meetings, design reviews, and pre-job briefings — risk becomes a shared awareness, not a hidden variable.

At its core, **risk assessment** is an act of imagination guided by evidence. It forces people to visualize failure in advance, to name the unlikely before it becomes unavoidable. It is not a paperwork exercise or a regulatory checkbox — it is leadership in analytical form. Mature organizations don't do risk assessments because they're required; they do them because they understand that every hazard identified early is a life protected later.

Readiness, on the other hand, is how organizations prove their risk analysis matters. A plan is only as strong as its rehearsal. Emergencies do not wait for comfort or convenience; they expose whether planning has become instinct. That's why drills, simulations, and debriefs are not interruptions to productivity — they are productivity in disguise. They build the muscle memory that allows people to respond with calm precision instead of fear.

The mature leader recognizes that readiness is not about predicting every scenario; it's about strengthening the decision-making process under stress. A well-trained, well-briefed team can adapt even when the script changes. They don't panic because they trust their system — and each other.

Section 6 closes the loop between foresight and action. It reminds us that every safe day was made possible by someone who imagined the unsafe day in advance and built protection against it. When safety culture reaches this level of maturity, risk isn't a threat — it's a teacher.

Leadership Truth:
Readiness is not luck; it's leadership rehearsed. The more an organization practices foresight, the fewer surprises it faces.

Day 21 – The Discipline of Foresight

The Moment

Every safe decision begins in imagination. The discipline of foresight is not about predicting the future — it's about preparing for it. Every procedure, design, and routine exists because someone once asked, *"What could go wrong?"* That question, simple as it seems, is the origin of every safety barrier we now take for granted.

In mature organizations, foresight isn't treated as pessimism; it's treated as stewardship. It's the mental discipline that refuses to confuse "hasn't happened" with "can't happen." The difference between luck and leadership is foresight — one waits for the unexpected, the other plans for it. The power of risk assessment lies in its honesty: it forces people to confront discomfort, to visualize failure without fear, and to learn from what hasn't yet occurred.

A good foresight exercise doesn't just identify hazards; it exposes assumptions. It reveals the hidden beliefs that quietly steer decision-making — beliefs like *"We've done it this way for years,"* or *"That's never been a problem before."* Those are the phrases that blind organizations to change. Foresight opens their eyes.

The Mindset

The discipline of foresight begins with humility. It accepts that systems are fragile and that experience, while valuable, can't see every angle. It's a mindset that invites diverse perspectives — operations, maintenance, engineering, safety, and even finance — into the same conversation because risk doesn't respect department boundaries. The purpose isn't to create fear but to create clarity.

Mature leaders teach their teams that risk assessment is not a task to be completed but a conversation to be continued. Every project, procedure, and new idea deserves a moment of structured curiosity — a pause to ask, *"What are we not seeing?"* The more freely people speak their concerns, the safer the system becomes. Foresight thrives in honesty and dies in hierarchy.

66

When foresight becomes habit, people stop waiting for danger to appear before they act. They think in contingencies. They question assumptions. They anticipate failure not as cynics, but as caretakers. That's what transforms safety from a policy into a practice — when people learn to see ahead, not react behind.

The Discipline

Foresight is strengthened through repetition. Risk assessments must be living documents, revisited as systems evolve. The disciplined leader schedules those reviews not because a regulation demands it, but because reality changes faster than memory. True foresight is not about avoiding all risk — it's about understanding it so completely that no risk takes the team by surprise.

Reflection

- Do we make time to imagine what could go wrong — before it does?
- Are our risk reviews living exercises or archived paperwork?
- How often do we challenge assumptions we've grown comfortable with?

Commitment

- I will lead with curiosity, not certainty.
- I will treat foresight as a daily discipline, not an annual event.

EBR Principle

Evidence-Based Safety treats anticipation as prevention. Seeing risk early isn't fear — it's foresight made visible.

Day 22 – When It's Not If, But When

The Moment

Preparedness is not a prediction of disaster — it's a declaration of care. The most resilient organizations don't waste time debating whether emergencies will happen; they ask how ready they'll be when they do. Every system eventually faces its test: a chemical leak, a medical emergency, a natural disaster, a power loss, or a human mistake. What happens next reveals whether safety is a belief or a behavior.

In those moments, panic is the product of surprise. Readiness, on the other hand, is the product of practice. It comes from countless drills, clear roles, and leadership that rehearses calm. The point of emergency planning isn't to script perfection — it's to build reflexes that make composure possible. People under stress don't rise to the occasion; they fall to their level of preparation.

The Mindset

Mature safety cultures understand that fear and readiness cannot coexist. Fear paralyzes; readiness empowers. The goal of preparedness is not to eliminate anxiety, but to replace it with confidence earned through repetition.

A well-built emergency plan does more than assign tasks — it builds trust. People know who will act, what will happen, and where to go. They believe in the plan because they've lived it, not just read it. The best leaders make preparedness personal: they treat drills as rehearsals for leadership, not interruptions to productivity.

Readiness is also psychological. It's about teaching teams that calm is contagious. When leaders move deliberately, speak clearly, and maintain presence, others borrow their composure. That steadiness turns chaos into coordination. Over time, the culture begins to understand that preparation is not paranoia — it's professionalism.

The Discipline

Preparedness becomes culture only through repetition. Emergency response must be reviewed, refreshed, and retrained regularly. Roles evolve, systems change, and new people join. The disciplined organization doesn't assume readiness — it verifies it. Each drill, after-action review, and simulation is a quiet investment in the system's future.

The proof of a mature safety culture isn't how it avoids emergencies; it's how gracefully it responds when they arrive.

Reflection

- Do we rehearse calm, or only hope for it?
- When emergencies happen, do people rely on memory or on muscle memory?
- Are our drills realistic enough to test readiness, not just compliance?

Commitment

- I will treat preparation as an act of leadership, not obligation.
- I will build confidence through rehearsal, not reassurance.

EBR Principle

Evidence-Based Safety defines readiness as repeatability under stress. What you practice calmly today becomes what you perform confidently tomorrow.

Day 23 – Reliability Through Routine

The Moment

Safety isn't sustained by inspiration — it's sustained by rhythm. The most reliable organizations don't rely on memory or mood; they rely on routine. Each checklist, inspection, meeting, and pre-job briefing is a heartbeat that keeps the system alive and aligned. Routines are not the enemy of flexibility — they're the foundation of it. They give people something solid to stand on when conditions change.

When a team begins to trust its routines, chaos loses its grip. Predictability creates calm. It allows people to act with focus rather than fear. A consistent routine doesn't just guide behavior — it communicates belief. It says, *"We care enough to do this right every time."* That message builds trust faster than any slogan ever could.

The danger comes when routines decay into rituals — when the actions remain, but the attention fades. Checklists become noise. Meetings become autopilot. The routine is still visible, but the meaning is gone. Mature leaders understand that repetition without reflection creates compliance, not commitment. Routine only builds reliability when it remains connected to purpose.

The Mindset

Routine is leadership made visible. It is how priorities become predictable and how culture becomes measurable. When leaders show up on time, start meetings with intent, and follow through on commitments, they're teaching through example that consistency is safety. People begin to internalize that safety isn't just what's written in policy — it's what's practiced in rhythm.

But discipline doesn't mean rigidity. The healthiest routines breathe. They evolve as systems and people evolve. The mindset of continuous improvement keeps repetition from becoming stagnation. The question isn't *"Are we doing this?"* — it's *"Is this still helping us stay safe?"* That small shift keeps routines alive.

Psychologically, routine provides something essential: **certainty**. The brain relaxes when it knows what comes next. SCARF theory calls this the need for predictability — the safety that comes from structure. When routines are reliable, people can focus on quality instead of anxiety. The mind that trusts its environment can think more clearly, communicate more honestly, and act more precisely.

The Discipline

Reliable systems don't run on motivation; they run on structure. Review your routines. Test them. Retire the ones that have lost meaning. Refresh the ones that still matter. Every repeated act is a message — make sure it's one worth sending.

Reflection

- Are our routines alive with meaning or empty with habit?
- Do people trust the system because it's consistent, or resent it because it's stale?
- What routine best demonstrates what we believe about safety?

Commitment

- I will honor the routines that protect people, not just repeat them.
- I will make consistency a visible act of leadership.

EBR Principle

Reliability is rhythm made visible. The strength of a safety culture isn't in its slogans — it's in the consistency of what people repeat with purpose every day.

Day 24 – Review: Practiced Readiness

The Moment

Readiness is not a document — it's a discipline. A safety culture becomes trustworthy when preparation turns from a checklist into a reflex. Emergencies, near misses, and system failures all test the same truth: whether the organization practices calm under pressure. In the end, what people do in a crisis is never new — it's a reflection of what they've been rehearsing all along.

True readiness isn't about memorizing steps; it's about understanding why they exist. It's built on layers of foresight, drills, communication, and trust. Each layer removes uncertainty until only clarity remains. The goal isn't to control every outcome but to ensure that when the unexpected happens, people still know what to do.

The best systems rehearse for stress. They practice small recoveries so that large ones aren't paralyzing. That's how psychological safety grows — not from slogans, but from confidence earned through repetition.

The Mindset

Readiness lives where awareness meets action. Mature leaders know that safety isn't the absence of danger — it's the presence of preparation. They treat every briefing, inspection, and exercise as an opportunity to strengthen reflexes, reinforce trust, and reduce panic.

This mindset recognizes that emergencies don't create character; they reveal it. When the alarm sounds, people don't invent new habits — they default to what leadership has modeled. Calm follows consistency. A team that sees its leaders steady and focused will follow their rhythm. That's why readiness is both technical and emotional: systems prepare procedures, but leadership prepares people.

When readiness becomes part of culture, anxiety decreases and confidence rises. People act with composure because they've practiced composure. They see preparation not as wasted time but as time invested — proof that leadership cares enough to plan ahead. Over time, that mindset transforms from compliance into pride.

The Discipline

Readiness fades without renewal. Drills, audits, and scenario reviews are not formalities; they're calibrations. The disciplined organization keeps its plans alive through testing and feedback. After every exercise, the question must be asked: *What did we learn?* — because improvement is the heartbeat of preparedness.

Reflection

- Do we treat drills as practice for perfection or preparation for composure?
- How often do we review readiness not as a policy but as a living behavior?
- When stress hits, do we default to panic or to process?

Commitment

- I will treat readiness as an act of care, not compliance.
- I will rehearse calm until it becomes part of who we are.

EBR Principle

Safety maturity is measured not by how few events occur but by how predictably people respond when they do. Readiness practiced is readiness proven.

SECTION 7 – INFLUENCE & CULTURE

Focus:

Role modeling, communication, and reinforcement — how leadership tone multiplies across systems.

Every culture has a sound. It lives in the way leaders respond to problems, the words they choose, and the consistency of their tone. Long before a team reads a policy or attends a training, they hear the culture — in meetings, in reactions, in how people talk about safety when no one official is listening. That sound becomes the rhythm others follow.

Influence is how leadership becomes culture. People may comply with what leaders require, but they will always imitate what leaders reveal. Behavior travels faster than instruction. When the most visible people on the floor — supervisors, operators, or senior leaders — model steadiness, curiosity, and care, those qualities echo through the system. The opposite is equally true. When shortcuts, sarcasm, or silence go unchecked, those behaviors spread even faster. The culture you have is the one you've trained through example.

Mature safety systems recognize that tone is a control point as real as any sensor or valve. The consistency of leadership behavior sets the emotional stability of the workforce. People don't just watch for direction — they watch for permission. If leaders can respond to near misses calmly, listen to bad news without blame, and reinforce accountability without humiliation, they teach that truth is safe here. That single message — *"It's safe to speak up"* — is the foundation of every high-performing safety culture.

Influence isn't about authority; it's about authenticity. Employees learn what's real by observing whether leaders mean what they say. A single contradiction between rhetoric and reaction can undo months of training. But when people see alignment — when the same leader who talks about safety also walks the site, listens first, and praises prevention instead of heroics — trust compounds. Culture coherence grows because credibility has evidence.

Reinforcement is where influence turns into permanence. Every compliment, correction, or comment adds another layer to the culture's memory. Over time, those layers become norms. That's why the most effective leaders don't just communicate — they curate behavior. They know every word teaches, every silence approves, and every reaction records a lesson.

A safe culture doesn't happen by mandate — it happens by mirror. The organization becomes a reflection of how its leaders behave under pressure, how they speak when they're tired, and how they handle the truth when it's uncomfortable. Influence is invisible until it echoes. And when it echoes with consistency, it becomes culture.

Day 25 – Everyone's Watching

The Moment

Every organization has cameras, but the ones that matter most are invisible — the eyes of those who watch what leadership does when no one else seems to be watching. Culture isn't built in meetings; it's built in moments. A supervisor pauses to put on gloves before entering a work area. A manager listens instead of reacting. An operator sees that pause and learns more in ten seconds than in any training course.

Everyone is watching, not because they're suspicious, but because that's how humans learn. We don't adopt standards — we imitate examples. The most powerful leadership messages are the ones people witness, not the ones they're told. A leader's smallest decision — to follow the rule, to check the permit, to speak calmly after an incident — sends a message about what's real here. The difference between what's written and what's witnessed defines the trust gap.

Safety culture is never invisible; it's either reinforced or undermined by every action leadership takes. Every time someone says, "We'll make an exception just this once," the system learns something. Every time someone says, "We'll do it the right way, even if it takes longer," the system learns something else. Those moments multiply — for better or worse — through imitation.

The Mindset

The mature leader understands that influence is constant. Whether they mean to or not, they are always teaching. Their tone, their urgency, their patience — all of it becomes the emotional climate others work within. Influence isn't about being perfect; it's about being consistent. When people see the same integrity in both comfort and crisis, they stop performing for approval and start acting from belief.

This mindset transforms leadership from enforcement to example. It recognizes that accountability spreads fastest through admiration, not authority. People copy what they respect. And respect isn't earned by speeches — it's earned by steadiness. When a leader shows the same care for a small risk as for a large one, or the same composure after an error as before a success, the team sees safety as a shared standard, not a selective one.

Everyone is a role model to someone. That realization doesn't create pressure — it creates purpose. Every choice becomes an opportunity to prove what the culture stands for.

The Discipline

Model what you expect. Praise what you want repeated. Correct what you can't afford to normalize. Leadership example is not a strategy; it's a system — one that runs on visibility and credibility.

Reflection

- What do my daily actions teach, even when I'm not speaking?
- Are people imitating my best habits — or my shortcuts?
- What would the culture look like if everyone acted the way I do when I'm under stress?

Commitment

- I will lead through example, not exception.
- I will make sure my actions teach what my words intend.

EBR Principle

Culture doesn't listen to what leaders say — it listens to what leaders repeat. Every action echoes, and eventually, the echo becomes identity.

Day 26 – The Echo of Leadership

The Moment

Words travel farther than footsteps. Long after a meeting ends or a decision is made, what people remember isn't the PowerPoint or the policy — it's how the message made them feel. The tone of leadership becomes the tone of the culture. If the sound of leadership is rushed, defensive, or dismissive, that rhythm will soon fill the hallways. If it's calm, thoughtful, and respectful, that rhythm will echo just as strongly.

Safety lives or dies in that echo. When a leader responds to a near miss with curiosity instead of anger, the message that travels isn't just about the incident — it's about psychological safety. When recognition is genuine and specific, people learn that their contribution matters. When words are careless, people learn that honesty is dangerous. Over time, the organization begins to sound like its leaders — not by instruction, but by imitation.

Leadership voice isn't just what's said; it's what's repeated. A leader who says "Safety first" but praises output over protection creates noise instead of harmony. The organization listens to what's amplified. That's why the echo of leadership is so powerful: it reveals the difference between values as spoken and values as lived.

The Mindset

Communication is not about volume — it's about resonance. The mature leader understands that clarity and consistency outlast charisma. They know that language shapes trust: every phrase either strengthens or weakens the bridge between intention and interpretation. Rhetoric is never neutral — it either builds confidence or confusion.

To lead through language, one must speak with precision and conviction. Precision creates understanding; conviction creates belief. Together, they create alignment. This mindset sees speech as design — every phrase, metaphor, and recognition moment becomes an instrument of culture. Words that match behavior amplify trust. Words that drift from reality create cynicism.

Great leaders don't speak to impress; they speak to reinforce. They repeat the right messages until the organization can repeat them back. They replace slogans with stories, because stories teach what systems feel like. Over time, those echoes harden into culture — a shared language of what "right" sounds like here.

The Discipline

Audit your words. Listen for what's being repeated on the floor or in the field. If the echo doesn't match your intent, change your tone, not just your message. Every phrase is an instrument of design — tune it until it plays truth.

Reflection

- What words or phrases do people repeat after hearing me?
- Does my tone create calm or pressure?
- When safety is discussed, do my words match my actions?

Commitment

- I will speak in a way that reinforces truth, not image.
- I will ensure that what I echo becomes worth repeating.

EBR Principle

The culture doesn't drift from silence — it drifts from mixed messages. Leadership language is the first alignment, and every echo defines what endures.

Day 27 – From Rules to Rituals

The Moment

Every system begins with a rule, but only becomes culture when that rule is remembered through ritual. A rule tells people what to do; a ritual reminds them why it matters. Rituals turn safety from an instruction into an instinct — from something we comply with to something we believe in. They give rhythm to reliability, and meaning to repetition.

Think about the small moments that happen every day — the pre-shift huddle, the safety walk, the two-second pause before starting a task. When done with purpose, these aren't chores; they're choreography. They keep the team aligned, focused, and connected. When they become mechanical or rushed, they lose their meaning — and culture begins to hollow out from the inside.

The health of a safety culture can often be measured by the sincerity of its rituals. Do people look forward to the daily briefing, or do they endure it? Do leaders use these moments to connect, or just to comply? The difference between obligation and ownership lives in those few minutes — the transition from talking about safety to living it.

The Mindset

Mature organizations understand that rituals aren't about routine — they're about reinforcement. They are how belief is practiced. A culture that repeats what matters keeps what matters alive. Every ritual, no matter how small, tells a story about values. A rushed ritual says, "This doesn't matter." A consistent one says, "This is who we are."

The mindset of ritual is intentionality. It's realizing that predictability isn't boredom; it's trust. When people can count on rhythm, they stop bracing for surprise and start contributing with confidence. The leader's role isn't to invent new rituals every week — it's to renew the meaning of the ones that exist. To pause, to reflect, to remind.

Rituals aren't limited to meetings or ceremonies. They live in gestures: the way a supervisor thanks someone for stopping a job, the way a team signs off a checklist together, the way silence is held after an incident. These moments teach continuity — that safety isn't a goal we reach, but a rhythm we maintain.

The Discipline

Protect the rhythm. Don't let ritual turn into routine. Refresh its meaning, reconnect it to belief, and make it visible. The moment a ritual loses emotion, the culture loses connection.

Reflection

- Which of our daily rituals still connect to purpose, and which have drifted into habit?
- Do I treat repeated moments as chores or as chances to reinforce belief?
- How do our routines make people feel — aligned or automated?

Commitment

- I will renew the meaning behind our safety rituals.
- I will treat repetition as rhythm — not redundancy.

EBR Principle

Culture is not what people believe once — it's what they practice together every day. Rituals are how belief becomes behavior.

Day 28 – Review: Cultural Coherence

The Moment

A strong safety culture doesn't happen by accident — it happens by alignment. Every part of the system, from leadership decisions to the smallest daily habits, has to speak the same language. When the words, actions, and symbols of a workplace move together, culture feels natural. When they move against each other, it feels forced.

Cultural coherence is what happens when everything fits — when the rules make sense, the rituals have meaning, and the rhetoric matches reality. It's the difference between compliance and conviction. Compliance does what it must; conviction does what it believes. You can tell which one your organization has by how it behaves when no one is watching.

Every message a leader gives, every role model people follow, every ritual they repeat — these are the instruments of coherence. If any of them play out of tune, the rhythm breaks down. That's why mature organizations spend less time adding new initiatives and more time tuning what already exists. They ask hard questions: Are our rules still true? Are our meetings still meaningful? Are our messages still believable? Alignment isn't a one-time task — it's a daily act of cultural maintenance.

The Mindset

The mature safety mindset is integrative, not additive. It doesn't chase the next slogan or tool; it focuses on harmony. When people understand how each element — Require, Reward, Reinforce, Ritual, Rhetoric, Role Model, and Routine — connects, they stop seeing safety as a checklist and start seeing it as a language.

This mindset values clarity over novelty. It recognizes that systems don't need more noise — they need coherence. When everyone speaks in the same rhythm of care, consistency, and accountability, the culture becomes self-sustaining. The goal is not perfection; it's alignment — an organization so attuned that safety feels like instinct, not instruction.

True coherence isn't about uniformity. It allows personality, creativity, and adaptation — but all within the same frame of belief. Just as a jazz band can improvise without losing the beat, a coherent culture allows flexibility without chaos. The rhythm of reliability holds it all together.

The Discipline

Examine the connections, not just the components. A culture's strength isn't in how many parts it has, but in how well they harmonize. Keep tuning the tone until the message, the method, and the meaning sound the same.

Reflection

- Do our actions sound like our words?
- Do our rituals reinforce what our role models display?
- Is our culture aligned enough that safety feels like second nature?

Commitment

- I will look for alignment before I look for innovation.
- I will protect the rhythm that makes our culture coherent.

EBR Principle

Safety culture isn't built by programs — it's built by harmony. When the system sings in one voice, trust becomes the soundtrack of work.

SECTION 8 – ALIGNMENT & MOMENTUM (ΔS)

Focus:

Measuring and managing the gap between intent and reality — the Delta of Safety.

Every culture lives in two versions: the one leaders describe, and the one employees experience. Between them lies a space — sometimes small, sometimes wide — where words, behaviors, and beliefs don't fully match. That space is ΔS: the Delta of Safety. It's the measure of how far reality drifts from intention, and how much energy the organization loses maintaining belief instead of building trust.

In engineering, Δ (delta) represents change or deviation from the ideal state. In culture, it represents the same thing — the difference between what's required and what's reinforced, between what leadership declares and what daily behavior delivers. A small delta means alignment: people hear the same message, see the same evidence, and trust the same process. A large delta means drift: what's written on the wall isn't what's lived in the work.

Safety systems don't fail overnight — they drift quietly. A shortcut tolerated once becomes a new standard. A missed follow-up becomes a pattern. A well-intended message becomes distorted through layers of interpretation. The danger isn't defiance; it's divergence. When what leaders say and what people see don't match, belief collapses under the weight of contradiction.

ΔS is not a verdict — it's a vital sign. Just like vibration or pressure tells the truth about a machine, cultural signals tell the truth about alignment. You can measure it in small ways: the lag between policy and practice, the tone of a morning briefing, the honesty in a debrief. Each one is data — evidence of where the culture is tight, and where it's stretching.

Managing ΔS requires humility. It's not about blame; it's about calibration. Mature organizations treat misalignment as feedback. They look at drift and ask, "Where did the signal fade? Where did silence replace reinforcement? What belief got lost in translation?" The goal is not to eliminate all variation — that's impossible. The goal is to make variation visible early, before it becomes failure.

Momentum depends on alignment. When leadership words and worker experiences move together, energy flows forward. People stop translating and start improving. Clarity replaces confusion. Trust replaces tension. That's the power of a small ΔS — it frees the organization to focus on growth instead of correction.

Safety excellence isn't perfection; it's precision. It's the constant act of tuning a living system to its truth — tightening every connection between belief and behavior. Because in the end, reliability isn't built by rules alone. It's built by reducing the distance between what we say and what we actually do.

Day 29 – The Gap Between Belief and Behavior

The Moment

Every safety system begins with belief — the conviction that harm is preventable, that risk is manageable, that people matter. But belief alone doesn't make a workplace safe; behavior does. The gap between those two — between what we say we believe and what we actually do — is where culture reveals itself. That's the safety delta, the measurable distance between intent and execution.

Drift rarely comes from defiance. It comes from quiet compromises made in the name of convenience: a skipped checklist, a rushed conversation, a near miss unreported because "it wasn't that bad." Over time, these small deviations accumulate like rust, weakening the integrity of both systems and trust. The danger isn't that people don't care — it's that they adapt faster than leadership corrects. When small misalignments go unaddressed, the organization begins to operate in two realities: the one in policy, and the one in practice.

The most dangerous phrase in safety is "We already do that." It assumes belief equals behavior, when in truth, they're rarely identical. Belief is intention. Behavior is evidence. And the role of leadership is to constantly reconcile the two.

The Mindset

The mature safety mindset understands that alignment is not an outcome — it's an ongoing act of awareness. Just as instruments need tuning, systems need calibration. A small gap today can become a wide delta tomorrow if it's ignored. Leaders who listen closely to the rhythm of the workplace — to the language, tone, and routines that shape daily life — can hear misalignment before it's visible in metrics.

Reducing ΔS begins with honesty. It requires the courage to see where our systems don't live up to our slogans, and the humility to treat drift as data, not disobedience. When leaders respond to inconsistency with curiosity instead of

86

control, people stop hiding the truth. They start revealing it. That transparency doesn't weaken authority — it strengthens integrity.

The power of small alignment gains cannot be overstated. Every time a leader follows through on a commitment, every time a team fixes a flaw instead of working around it, the delta shrinks. Trust grows. Consistency becomes contagious.

The Discipline

Watch for drift early. Listen for it in excuses, in workarounds, in silence. The sooner you detect deviation, the sooner you can restore integrity. Tighten the loop between belief and behavior daily — not with punishment, but with presence.

Reflection

- Where do my words and my actions disagree most?
- What small tolerances have become cultural norms?
- How do I make it safe for others to tell me where drift exists?

Commitment

- I will measure belief by behavior, not assumption.
- I will treat drift as feedback, not failure.

EBR Principle

Safety alignment isn't declared; it's demonstrated. Every action that matches belief turns culture from promise into proof.

Day 30 – Closing the Loop

The Moment

Every safety journey ends where it began — in human choice. Systems may guide it, policies may sustain it, but progress only endures when people decide to close the loop between knowing and doing. The most mature organizations understand that the true finish line isn't compliance; it's continuity. Safety doesn't reset each year with new goals or slogans — it deepens through cycles of learning, alignment, and renewal.

Closing the loop means refusing to let insight die in isolation. Every audit, every investigation, every moment of reflection is wasted if it doesn't result in visible change. Too often, lessons learned are documented but never applied. The feedback gets captured, but not circulated. The system "knows," but the people never feel the difference. When learning doesn't loop back into leadership behavior, the organization becomes fluent in reflection but illiterate in action.

The loop begins with detection, but it ends with discipline. Each discovery — a near miss, a weak signal, a drift in standards — is an invitation to strengthen the system. The work isn't glamorous; it's repetitive, steady, and often invisible. But that repetition is what makes reliability real. Every loop closed reduces ΔS — the distance between belief and behavior, between policy and practice.

The Mindset

Closing the loop requires humility and persistence. It asks leaders to be stewards, not saviors — to see safety not as an event to celebrate, but as a system to care for. The mature mindset understands that improvement isn't a burst of innovation; it's the compounding effect of small, consistent corrections.

Feedback loops are acts of respect. They prove that every observation mattered, every voice was heard, and every insight had impact. They tell the workforce, "We listened — and we acted." That connection is what builds cultural trust. Without it, even the best-intentioned systems breed cynicism.

To close the loop, leaders must do three things consistently: listen, act, and report back. Listening captures truth. Acting turns it into improvement. Reporting back completes the circuit — transforming accountability from a one-way demand into a shared dialogue. The result isn't perfection, but coherence.

The Discipline

Never let learning stop at awareness. Trace it to action. Track it to outcome. Teach it forward. A lesson unshared is a risk unremoved. Capture what happened, what changed, and who made it better — then make that knowledge visible.

Convert reflection into improvement plans with clear ownership and follow-through. Verify completion, document the results, and use them to strengthen the system. Every lesson only fulfills its purpose when it changes behavior — when what was learned becomes what's lived.

Reflection

- Where in our system does learning stall before it becomes change?
- Do people see their feedback result in visible improvement?
- What loops can I close this week to turn reflection into renewal?

Commitment

- I will make learning visible by turning every insight into improvement.
- I will protect the integrity of our system by completing every feedback loop.

EBR Principle

Sustained safety isn't built by catching every failure — it's built by closing every loop. Reliability grows stronger each time truth returns home as change.

Bonus Day – The Safety Equation

Reliability ≈ 1 / ΔS
The tighter the alignment between what's required and what's reinforced, the safer the system becomes.

The Moment

Every organization has a formula, whether it knows it or not. It's written not in policies or posters, but in patterns — how decisions are made, how feedback is received, and how truth travels through the system. The Safety Equation is the simplest way to describe that reality: the smaller the gap between what we *say* and what we *do*, the greater our safety, trust, and reliability.

ΔS — Delta Safety — represents the distance between leadership's intention and the organization's execution. It measures drift: that slow, invisible slide between ideal and actual, between written standards and lived behavior. Every time a requirement is unclear, a rule inconsistently applied, or a feedback loop left open, ΔS widens. And as it does, confidence erodes. People begin to adapt around inconsistency, and culture turns reactive.

Safety isn't built by slogans about zero harm; it's built by shrinking that delta — by closing the space between belief and behavior. That's what maturity looks like: not perfection, but alignment.

The Mindset

The Safety Equation reminds us that systems don't fail all at once — they fail in fragments. A missed inspection here, an unreported near miss there, a shortcut tolerated because it "worked." Each moment adds just a fraction to ΔS. Over time, those fractions compound into fragility.

Leaders who understand this equation stop chasing outcomes and start managing alignment. They study the space between words and actions. They ask, "What do our people experience as normal?" and "Does that match what we've declared as right?" This curiosity is not criticism — it's calibration.

When ΔS shrinks, culture becomes self-correcting. People speak up earlier because they trust the reaction. Standards hold because leaders reinforce them consistently. Learning loops close faster, and improvement becomes continuous instead of conditional. Safety stops being an initiative and starts being an instinct.

The Discipline

Treat ΔS the same way you would treat any critical variable — monitor it, trend it, and control it. Use reflection, observation, and feedback to measure where intention drifts from reality. Correct small misalignments before they become habits.

Consistency is the real leading indicator. The tighter your alignment, the less energy is wasted managing belief and the more is invested in building trust.

Reflection

- Where do our words and systems disagree most often?
- What parts of our culture rely on personality instead of process?
- How small can we make our ΔS — and how will we know?

Commitment

- I will lead by reducing the distance between what we require and what we reinforce.
- I will measure progress not by what's promised, but by what's practiced.

EBR Principle

Safety is not the absence of accidents — it's the absence of drift. Alignment is the most powerful control in any system, because when belief and behavior move as one, reliability becomes inevitable.

BONUS Section

How to Use the Bonus Section

The 30-day journey was designed to shift how you *think*.
This Bonus Section exists to help you shift how you *act*.

These final tools are not appendices — they're extensions of practice. Each one turns reflection into rhythm, giving you a way to see your thinking at work in the real world. The purpose isn't to add more reading, but to give you structured places to pause, connect, and apply.

Use these pages the same way you use a mirror: not to admire, but to adjust. Return to them at the end of each project, safety meeting, or quarter and ask, *What have we aligned? What have we drifted from?*
They're meant to be written in, debated, shared, and re-read — because repetition is where Mindsets become methods.

You'll find frameworks for reflection, self-audits, and planning tools that link the human side of safety (belief, tone, leadership) with the system side (structure, standards, accountability).

The daily readings built awareness.
This section builds alignment.

Use it as your maintenance manual for culture — not to preserve perfection, but to sustain direction.
Because the goal isn't to finish the 30 days — it's to keep them alive.

The Safety Equation

The Safety Equation – Deep Dive & Reflection Tool

Every culture has a *Delta*, a space between what leaders say and what people see. In technical systems, that gap shows up as process variation. In human systems, it shows up as drift — a slow, quiet slide from intention to interpretation. The smaller the drift, the safer the system. The larger it grows, the more we rely on luck instead of discipline.

That relationship can be expressed as a simple equation:

Reliability $\approx 1 / \Delta S$

Where ΔS (Delta Safety) represents the difference between what's *required* and what's *reinforced*.

When ΔS is large, you are managing two systems at once — the one you believe you're running, and the one people are actually living. The two are out of phase. Policies say one thing, practices say another. People compensate, adapt, and develop shadow systems that make sense locally but undermine control globally. These organizations often look successful from a distance but feel chaotic up close.

When ΔS is small, belief and behavior align. The culture becomes predictable. Communication becomes honest. Systems become self-correcting. That's the heart of safety maturity — not control for control's sake, but consistency for trust's sake.

A small ΔS means leaders reinforce what they require, reward what they reinforce, and routinely model what they reward. Alignment becomes rhythm.

Reflection: Measuring Your ΔS

Ask yourself the following six questions.

Question #1: Belief vs. Behavior Do my words match what people experience daily?

Belief vs. Behavior – The True Test of Safety Culture
Every organization has two operating systems: the one people **believe in** and the one they actually **behave in**. Beliefs live in words — mission statements, policies, and posters. Behavior lives in choices — what people do when no one's watching, how they react under pressure, and what they tolerate when standards are inconvenient. The distance between those two worlds defines culture. The tighter they align, the safer the system becomes.

Belief without behavior is philosophy. Behavior without belief is compliance. Both can exist temporarily, but neither sustains safety for long. Real safety happens when what people say they value is indistinguishable from what they routinely demonstrate. That alignment doesn't happen by accident — it's built through leadership consistency, visible reinforcement, and the discipline to hold the line even when production pressures rise.

Most organizations don't suffer from lack of belief; they suffer from lack of follow-through. Everyone agrees that safety is important — until a deadline slips, a shipment is late, or a supervisor is measured by numbers alone. In those moments, culture reveals itself. People watch what leaders do, not what they declare. When leadership protects a standard even at the cost of convenience, belief hardens into trust. When leadership bends a rule in silence, behavior rewrites belief. Every exception tells the organization what truly matters.

Closing the gap between belief and behavior requires courage, not charisma. It means confronting drift early, asking uncomfortable questions, and rewarding the quiet consistency that often goes unnoticed. It means creating systems where safety isn't left to memory or motivation — it's built into how work happens. The most credible leaders aren't the loudest about safety; they're the most predictable about it. Their behavior *is* their belief.

EBR Principle:
Belief sets direction; behavior proves it. The closer they align, the less energy is wasted managing compliance — because the system itself becomes the teacher.

Question #2: Policy vs. Practice Are our standards enforced equally, or situationally?

Policy vs. Practice – The Proof of What We Really Value

Policies tell the story leadership wants to believe; practice tells the story everyone else already knows. The space between them — however small — defines credibility. Policies exist to provide clarity. They define intent, boundaries, and expectations. But without practice, they remain theory — words preserved in binders while reality unfolds in shortcuts, improvisations, and workarounds. Every time a procedure is skipped with a wink or a rule is bent without consequence, the organization teaches something unintended: that policy is optional when it's inconvenient. Over time, that erosion of discipline becomes invisible. "How we actually do it" quietly replaces "how we're supposed to do it," and drift becomes the new normal.

If compliance requires heroics, the process—not the people—is broken. A well-designed policy supports reality instead of denying it. It fits the rhythm of work, respects human limits, and aligns with the pressures employees actually face. When policy feels practical, it becomes part of identity, not just instruction.

The strongest cultures don't manage compliance; they design for consistency. Supervisors model the policy in action. Procedures are reviewed with the people who perform them, not just approved by those who write them. Violations are studied for insight, not ignored for convenience. Over time, practice begins to mirror policy — not because enforcement got stricter, but because belief got clearer. People stop treating the rules as external control and start treating them as internal discipline.

The ultimate test of leadership integrity is whether the written word survives contact with the real world. When it does, culture gains coherence. When it doesn't, safety becomes performative.

EBR Principle:

Policy defines what we intend to value. Practice reveals what we actually value. The smaller the gap between the two, the safer — and more believable — the system becomes.

Question #3: Message vs. Method Does our tone change under pressure?

Message vs. Method – When Communication Becomes Culture

Every organization communicates two messages: the one it *intends* to send and the one its *methods* actually deliver. The difference between them defines whether communication builds clarity or breeds confusion. A powerful message loses all meaning when the method contradicts it. Tone, timing, and consistency speak louder than slogans ever will.

A leader can say, "Safety first," yet schedule production overtime that forces fatigue. They can preach "open communication," yet react defensively to bad news. In those moments, the method cancels the message. People stop listening to words and start learning from behavior. Over time, they begin to treat leadership messages as signals of risk rather than signals of trust — reading the space between the lines to decide what's truly safe to say or do. The culture learns to decode tone, not intent.

The alignment of message and method begins with self-awareness. Communication is more than what's said; it's *how* and *when* it's said. Urgency without empathy feels like pressure. Praise without explanation feels like favoritism. Correction without consistency feels like threat. The mature communicator knows that the delivery system is as important as the content itself — because emotion always arrives before logic.

In safety, the cost of misaligned communication is measured in silence. Teams that don't trust the method stop sharing the truth. Near misses go unreported. Problems stay local until they become global. Yet when leaders match their message with method — calm tone, predictable follow-up, transparent reasoning — they make honesty feel safe. People begin to trust that what's said will be supported by what's done.

EBR Principle:
Message defines intention. Method defines integrity. When both align, communication stops being instruction and starts becoming evidence.

Question #4: Reward vs. Requirement: Who receives praise — the planner or the firefighter?

Reward vs. Requirement – The Real Teacher of Culture

Every organization has two scoreboards: what it requires and what it rewards. The difference between them teaches the truth about values. Requirements tell people what leadership says matters. Rewards tell them what really matters. When those two signals align, culture accelerates. When they diverge, credibility collapses.

A policy may say, "Follow the procedure," but if the fastest worker gets the praise — even when they cut corners — the lesson is clear: speed matters more than safety. Requirements shape compliance; rewards shape belief. People don't imitate rules; they imitate results. Whatever earns recognition or applause becomes the real operating standard. Over time, that standard rewrites the system — quietly, powerfully, and often without a single word spoken.

Reward is the most honest feedback loop in any organization. It reinforces not just what gets done, but how it gets done. If leadership wants a culture of safety, it must celebrate the quiet behaviors that make incidents impossible — not just the recoveries after something goes wrong. The truest recognition often belongs to the ones who kept the problem from ever appearing.

Alignment between reward and requirement is a sign of integrity. It means the system's incentives no longer compete with its values. The same actions that earn approval also uphold safety. When that happens, accountability stops feeling like enforcement and starts feeling like belonging. People begin to internalize the standard, not because they have to — but because it's who they are.

Every organization gets more of what it rewards and less of what it ignores.

EBR Principle:

Requirements define the rules of the game. Rewards decide how it's played. When both align, compliance becomes conviction — and belief becomes behavior.

Question #5: Ritual vs. Reality Are our routines living proof of our values, or just repetition?

Ritual vs. Reality – The Test of Authentic Culture

Every culture has its rituals — meetings, safety moments, recognition events, or kickoff huddles — but not every culture connects those rituals to reality. Rituals are supposed to remind people of purpose, yet when purpose fades, they become choreography without conviction. What was once meaningful becomes mechanical. People still show up, but their hearts don't.

The gap between ritual and reality is one of the quietest indicators of cultural decay. When the ritual says "we care," but the follow-through says "we're busy," trust erodes. When leadership opens meetings with talk of safety but ends them with production demands, the ritual becomes hollow. People stop believing what they hear and start believing what they see.

That doesn't mean rituals are bad. In fact, they are vital — the rhythm that holds a system together when pressure mounts. But for rituals to work, they must be connected to behavior and belief. A pre-job safety talk that sparks reflection is powerful. A checklist read by rote is not. A recognition moment that tells a real story reinforces values. A scripted applause for convenience undermines them.

Leaders can close the gap between ritual and reality by restoring meaning. Ask, *What belief does this express? What behavior does it reinforce? What emotion does it leave behind?* If the answer isn't clear, the ritual needs renewal — not removal. Rituals should evolve as understanding deepens, always pointing back to truth.

Because the strongest cultures don't worship routine — they honor relevance. Their rituals aren't performance; they're presence. They remind people not just what to do, but why it matters.

EBR Principle:
Rituals lose power when they stop reflecting belief. Reality restores it when leaders reconnect rhythm to meaning.

The Leadership Alignment Check (R^3/R^4 Lens)

Behind every strong safety system is a leadership system — a network of visible and invisible cues that tell people what really matters. The R^3/R^4 Model explains this dynamic:

R^3: Require, Reward, Reinforce.
These are the levers of leadership — the things we *say* and *do* to shape direction.

R^4: Rituals, Rhetoric, Role Models, Routines.
These are the mirrors of culture — the things the organization *repeats* until they become normal.

Alignment occurs when **R^3 and R^4** move in rhythm.
Misalignment occurs when they compete.

You can tell instantly when they're out of sync. Leaders may require safety, but reward speed. They may reinforce accountability, but tolerate silence. They may preach teamwork, but role model isolation. The organization quickly learns which message actually governs behavior.

To realign, start by checking for coherence:

Leadership Alignment Check

Require
- *Reflection Question:* Are our expectations written, visible, and stable over time?
- *Alignment Signal:* Consistent clarity in decisions and messaging

Reward
- *Reflection Question:* Do we publicly recognize behaviors that reflect values, not just results?
- *Alignment Signal:* People emulate the right examples

Reinforce
- *Reflection Question:* Are our corrections calm, consistent, and fair?
- *Alignment Signal:* Feedback feels safe and predictable

Rituals
- *Reflection Question:* What rhythms prove our priorities each week?
- *Alignment Signal:* Meetings, huddles, and debriefs mirror intent

Rhetoric
- *Reflection Question:* Does our language produce trust or tension?
- *Alignment Signal:* Teams use "we" more than "they"

Role Models
- *Reflection Question:* Who do people really admire here — and why?
- *Alignment Signal:* Admiration flows toward aligned behavior

Routines
- *Reflection Question:* Are our daily habits teaching consistency or chaos?
- *Alignment Signal:* Predictability across shifts, teams, and leaders

Alignment is not perfection. It's pattern. Every conversation, every meeting, every reaction teaches what this organization truly values.

When you require what you reward, reinforce what you require, and reflect it in daily routines, ΔS tightens. The system becomes believable. People start to trust that what is *expected* is what will actually be *experienced*. That's what culture coherence looks like — leadership integrity translated into daily rhythm.

Reflection

- What part of our R^3 system feels the strongest right now?
- Where does R^4 feel disconnected from it?
- How could one visible change — in routine, recognition, or language — help close that gap?

The Seven Anchors of Safety Culture

Every strong safety culture rests on a few immovable anchors — principles that hold the system steady when conditions shift. These anchors aren't slogans or policies; they're disciplines of thought and behavior that keep people aligned even when pressure rises. They balance the human and the technical, the emotional and the procedural. Each one represents a choice: to act with clarity instead of confusion, discipline instead of drift, curiosity instead of complacency.

When these anchors are practiced together, safety becomes more than a program — it becomes a pattern. They are the rhythm beneath reliability, the quiet habits that keep systems stable and people safe.

The Seven Anchors of Safety Culture:

1. Control
2. Clarity
3. Competence
4. Curiosity
5. Change
6. Courage
7. Consistency

EBR Principle: Stability isn't accidental — it's anchored by behavior that stays true under pressure.

Control – Management Commitment, Policies, Accountability

Control is not about command; it's about containment — keeping chaos inside boundaries where it can't harm people or progress. In safety, control begins with leadership commitment. When management visibly owns outcomes, accountability becomes cultural rather than conditional. Policies then serve as the architecture of that control — written clarity that turns good intent into consistent behavior. A strong policy doesn't just say what's forbidden; it explains what's protected and why.

But control without accountability is paperwork. The real measure of control is whether expectations hold under pressure — whether a leader will choose the policy over convenience, even when nobody's watching. True accountability doesn't seek fault; it seeks fidelity. It ensures that every deviation is examined, not ignored, because boundaries protect more than compliance — they protect people.

When control is mature, it doesn't feel oppressive; it feels secure. People work confidently because they know what "good" looks like and trust that leadership means what it says. Safety, then, isn't the absence of risk; it's the presence of reliable control — built through commitment, codified through policy, and sustained through accountability.

EBR Principle: Control without consistency creates fear; control with clarity creates freedom.

Clarity – Communication, Structure, Standards

Clarity is the bridge between leadership's intent and the workforce's reality. Without it, even the best systems drift into confusion. Every great safety culture depends on two things: clear communication and a structure that channels it. When messages travel cleanly through well-defined roles, understanding replaces interpretation. Ambiguity disappears, and with it, frustration.

Structure gives safety its shape — a visible hierarchy of purpose that makes accountability possible. It ensures that everyone knows their lane, their limits, and their leverage. But structure alone isn't enough; standards must translate structure into behavior. Standards define the expected rhythm of excellence — not to restrict creativity, but to create alignment.

Clarity doesn't mean over-explaining; it means eliminating guesswork. When expectations are known, communication becomes proactive instead of reactive. People can plan instead of scramble. The organization stops relying on personality and starts relying on process.

A clear system doesn't shout to be heard; it hums with predictable rhythm. And in that rhythm, trust grows — because people see that words match actions, and structure supports both.

EBR Principle: Confusion is the enemy of control. Clarity transforms intention into impact.

Competence – Training, Procedures, Integrity

Competence is the invisible confidence that keeps a system calm under stress. It begins with training — not as a compliance exercise, but as a transfer of judgment. Great training doesn't just teach steps; it teaches thinking. It builds instincts strong enough to guide people when the procedure isn't in front of them.

Procedures, then, serve as the scaffolding of that competence. They capture what experience has proven to work and prevent the reinvention of risk. When written clearly and practiced routinely, they create reliability through repetition. But procedures alone can't replace integrity. Integrity ensures that people follow the standard even when shortcuts seem faster — that consistency wins over convenience.

Competence is not the absence of error; it's the presence of readiness. It's what allows teams to stay steady when complexity rises. When skills are honed, standards known, and values intact, people make fewer mistakes — and recover from them faster.

EBR Principle: True competence isn't memorized; it's embodied. It shows up when no one's watching.

Curiosity – Audits, Observation, Investigation

Curiosity is the engine of every improvement. It's what turns compliance into learning. In mature safety systems, audits, observations, and investigations aren't acts of judgment — they're acts of curiosity. They ask, "What is this system trying to tell us?" instead of "Who failed?"

Audits create visibility. They show what the process produces, not just what people perform. Observation adds humanity — it watches how work really happens, not how the procedure imagines it. Investigation completes the cycle by connecting cause to context, helping teams see patterns instead of blame.

Curiosity thrives where leadership listens more than it lectures. It replaces fear with inquiry and turns near-misses into near-masters — moments where understanding expands faster than punishment spreads. In such a culture, information flows upward freely, because people know that truth is safe to share.

Curiosity is how safety stays alive. It keeps learning faster than failure can accumulate.

EBR Principle: Judgment stops progress; curiosity starts it.

Change – MOC, Quality Assurance, Adaptability

Change is the ultimate stress test of a safety system. Management of Change (MOC) ensures that new ideas don't outpace wisdom. It forces pause before progress — inviting the right people with the right expertise to ask the right questions. MOC isn't bureaucracy; it's protection from unintended consequences.

Quality assurance adds another safeguard — confirming that what was built or modified still meets the standard. It's how the organization proves that safety isn't situational, even under innovation. And adaptability ties both together: the ability to evolve without eroding control.

A healthy safety culture welcomes change, but never blind change. It knows that progress without precaution can turn optimism into accident. Adaptability means integrating lessons without losing discipline — growing stronger, not looser, with every iteration.

When MOC and QA are respected, change becomes an opportunity to demonstrate maturity, not a reason to gamble with risk.

EBR Principle: Change without control is chaos. Adaptability is the art of safe evolution.

Courage – Risk, Preparedness, Leadership Integrity

Courage is the quiet force behind every act of prevention. It takes courage to stop the job, to challenge assumptions, to speak truth to authority. In safety, courage is not dramatic — it's disciplined. It shows up as the willingness to face risk honestly and prepare relentlessly.

Preparedness is courage in advance. It's the choice to imagine what could go wrong and act before it does. The mature leader rehearses risk, not to create fear, but to build calm under pressure. That's what integrity looks like in motion

106

— making the safe decision even when it's inconvenient, even when nobody would know otherwise.

Courage aligns belief with behavior. It bridges the gap between policy and practice, turning values into visible choices. When leaders demonstrate integrity under stress, they don't just prevent incidents — they teach bravery by example.

EBR Principle: Courage isn't the absence of fear; it's the presence of alignment between conviction and action.

Consistency – Rituals, Rewards, Reinforcement

Consistency is the heartbeat of reliability. It's what turns principles into patterns and patterns into culture. Rituals — those repeated behaviors like pre-shift meetings, debriefs, and recognition moments — give safety its rhythm. They remind people not only what to do, but why it matters.

Rewards then amplify that rhythm by signaling which behaviors deserve repetition. When recognition celebrates prevention instead of rescue, culture shifts from reaction to reliability. Reinforcement sustains the cycle — leaders follow up, repeat, and re-model until the right habits feel natural.

Consistency doesn't mean rigidity. It means predictability — the comfort of knowing that the rules don't change with the room. When leadership tone, standards, and actions remain stable, people relax into focus. Chaos fades, confidence grows, and the system becomes self-correcting.

EBR Principle: Consistency is credibility in motion. What leadership repeats, people remember.

The 10 Reflections of a Safety-Minded Leader

Safety maturity isn't defined by how loudly someone talks about safety — it's revealed in how quietly they practice it.

These reflections are not rules but reminders — subtle calibrations for the leader who understands that culture shifts first within the individual.

Each line is both a mirror and a measure: an invitation to notice how your own habits teach others what "safe" really means.

A mature safety leader...

1. ...speaks last and listens first.
2. ...responds to truth with curiosity, not control.
3. ...writes down lessons so the next person doesn't have to relearn them.
4. ...treats observation as investment, not inspection.
5. ...knows that the tone of correction is part of the correction.
6. ...follows the same rules they enforce — especially when it's inconvenient.
7. ...turns accountability into encouragement, not embarrassment.
8. ...treats every deviation as data, not defiance.
9. ...balances authority with humility and policy with patience.
10. ...understands that every word, pause, and reaction teaches the culture what leadership really values.

EBR Principle:
The mindset of a leader becomes the method of the team.

From Awareness to Action Worksheet

Insight without action fades fast. The goal of this 30-day journey isn't just to think differently — it's to *lead differently*.

This worksheet helps close the gap between awareness and behavior, between what you've learned and what you'll live.

Use it after finishing the book — or anytime a new insight demands structure. Each question moves you from thought to traction, transforming belief into visible behavior and behavior into sustainable system change.

Biggest Mindset Shift

What idea changed how I see safety?

Capture the one insight that disrupted your old way of thinking — the truth that won't let you go back to "normal."

Behavior to Practice

What action will prove that shift?

Define one behavior that turns belief into evidence. Be specific, measurable, and visible to others.

System Change Needed

What structure or process will sustain it?

Identify what must change beyond yourself — a routine, meeting, checklist, or decision path that will reinforce this new standard.

Measure of Success

What will I see or feel when this becomes normal?

Describe the signals of alignment — smoother operations, calmer reactions, greater trust. When this mindset becomes habit, how will the culture sound, look, and feel?

EBR Principle:

Change doesn't stick because people believe in it — it sticks because systems reinforce it.

The 12 Rules of Safety

(As taught by the 30-Day Safety Mindset)

1. **Safety begins with belief, not bureaucracy.**
   The strongest systems start in the mind — where accountability becomes personal and commitment becomes visible. Until people believe safety is a value, not a rule, every policy will remain optional in practice.

2. **Control comes from clarity, not control.**
   The safest organizations don't micromanage; they make expectations unmistakable and systems trustworthy. Clear purpose reduces the need for pressure — people perform best when they know exactly what "good" looks like.

3. **What you tolerate, you teach.**
   Every unchallenged shortcut rewrites the standard. Every silence resets the rule. The culture will always drift toward whatever leaders allow, not what they announce.

4. **People support what they help create.**
   Involvement isn't a courtesy — it's a control measure. Shared ownership builds safer outcomes. When workers design the system, they defend it; when they're excluded, they work around it.

5. **Procedures protect freedom.**
   Discipline is not restriction; it's the structure that allows initiative without chaos. A written standard gives people confidence to act decisively within safe boundaries.

6. **Learning is the heartbeat of prevention.**
   Every audit, observation, and near miss is a tuition payment for wisdom — if we're humble enough to collect it. A culture that documents lessons builds memory; one that doesn't is condemned to repeat mistakes.

7. **Change is the ultimate test of culture.**
   When the system flexes without breaking, you know safety is real — not rehearsed. Maturity is proven not by how we perform in routine, but how we adapt under uncertainty.

8. **Fear hides truth; curiosity reveals it.**
   A culture that punishes mistakes will repeat them. A culture that studies them will outgrow them. Psychological safety is not softness — it's the structure that lets hard truths surface early.

9. **Preparedness is leadership made visible.**
   Readiness isn't built in emergencies — it's built in everyday repetition, ritual, and rhythm. Every drill, checklist, and review is a rehearsal of reliability, not a compliance exercise.

10. **Rituals remember what policy forgets.**
    The habits we repeat teach louder than the memos we write. A small, consistent action done with purpose will shape belief faster than a thousand slogans.

11. **Reward what reflects your values, not just what achieves your goals.**
    The applause you give today becomes the behavior you get tomorrow. Recognition that honors integrity builds trust; recognition that honors shortcuts builds risk.

12. **Reliability is the reward of alignment.**
    When belief, behavior, and system design all agree — safety stops being a program and starts being a property of the culture. The tighter the alignment between what's required and what's reinforced, the less energy leadership wastes managing belief.

EBR Principle:
Safety excellence isn't achieved by rule or ritual alone — it's achieved when leadership integrity, system design, and human behavior move in the same direction.

The Closing Message — Safety as Alignment

Safety is not a department.
It's not a scorecard, a slogan, or a set of posters on the wall.
Safety is a discipline of alignment — the daily act of bringing belief, behavior, and system design into agreement. When those three align, safety stops feeling like effort and starts feeling like culture. It becomes the quiet confidence of knowing that what we say, what we do, and what we allow all point in the same direction.

The goal has never been control — the goal is coherence.
Control demands compliance; coherence invites commitment. Control can make people follow rules while resenting them; coherence makes them care about the rules because they make sense. A coherent system doesn't fight human nature — it partners with it. It gives people the clarity to act safely without waiting for permission, because the standard is already written into the rhythm of work.

Every decision teaches the next one how to behave.
When leadership reacts with fairness instead of frustration, the organization learns calm. When a supervisor enforces a standard with steadiness instead of sarcasm, the team learns respect. When someone admits a near miss without fear, everyone learns that truth is safe here. These are the moments where alignment is either reinforced or fractured. Systems don't drift by accident — they drift through small inconsistencies repeated often enough to become invisible.

Alignment is not perfection; it's pursuit. It's the continuous maintenance of integrity between message and method. That's why safety is a leadership behavior before it's an organizational outcome. The way you plan a job, open a meeting, review a report, or respond to feedback either tightens or widens the gap between belief and behavior. The work is not about eliminating all risk; it's about eliminating all confusion. People can manage risk — what wears them down is uncertainty.

When safety becomes alignment, accountability feels fair, procedures feel useful, and leadership feels believable. Conversations become clearer. Inspections become learning instead of judgment. Performance becomes predictable not because people fear mistakes, but because the system no longer leaves them guessing. The most reliable organizations aren't the ones that chase zero incidents; they're the ones that refuse to tolerate zero learning.

Alignment is the invisible architecture of trust. It's what allows a team to move fast without losing control, to adapt without drifting, and to correct without blaming. It's what transforms compliance into competence and turns culture from a campaign into a living system.

Safety isn't something you do once; it's something you become through repetition, reflection, and relentless honesty. Each day, you align the visible and the invisible — the structure of the system with the spirit of the people who live inside it.

Because what you align today becomes the safety someone else inherits tomorrow.

About the Author

Andy E. Page, Jr., Ph.D.

Founder, **EBR Technologies**
Creator of the **Evidence-Based Reliability (EBR)™** and **RCM-FX™**
frameworks

Andy Page is a reliability engineer, strategist, and educator who has spent more than two decades helping industrial organizations transform the way they think about maintenance, performance, and culture. His work bridges two worlds — the precision of data and the discipline of leadership.

As the founder of **EBR Technologies**, Andy developed the Evidence-Based Reliability (EBR) framework, a practical approach that helps teams replace emotion with evidence and chaos with control. His **RCM-FX** method redefines classical reliability-centered maintenance with deeper categorization of failure effects, layered protection logic, and a culture-first mindset that connects the shop floor to the boardroom.

Over his career, Andy has guided clients across manufacturing, utilities, energy, and consumer goods — helping leaders and technicians alike build systems that think before they break. His teaching style combines technical clarity with cultural insight, making reliability not just a technical function, but a leadership behavior.

When he's not writing or consulting, Andy speaks to global audiences about the intersection of foresight, data, and discipline — and how evidence can become the most trusted voice in an organization.

About EBR Technologies

EBR Technologies (Evidence-Based Reliability) is a reliability consulting and training organization focused on helping clients build systems that think, plan, and act with discipline.

Founded on the belief that reliability isn't assumed — it's engineered, EBR Technologies equips organizations with tools and frameworks to:

- Engineer foresight through structured analysis and evidence-driven planning.
- Strengthen execution through Work Execution Management (WEM) systems that eliminate friction.
- Shape culture through the R^3/R^4 Model — aligning what leaders Require, Reward, and Reinforce with what the organization's Rituals, Rhetoric, Role Models, and Routines display.

EBR's work spans reliability improvement roadmaps, criticality analysis, PM optimization, asset walkdowns, and full-scale cultural alignment programs designed to make evidence the language of leadership.

EBR Technologies
Evidence is our authority.

www.ebrtechnologies.com
info@ebrtechnologies.com

Author's Note on the Use of AI

This book was written in collaboration with an artificial intelligence tool — not as a shortcut, but as a companion in reflection.

Every lesson, mindset, and maxim within these pages originates from my years of teaching, consulting, and field experience in safety, reliability, and culture. The principles draw from my established models — the R3/R4 Culture Framework, the Evidence-Based Thinking philosophy, and the broader discipline of Leadership Alignment that I've practiced and refined across industries and organizations.

AI served here as an instrument, not an author. Like a disciplined editor with infinite patience, it helped shape language, surface clarity, and maintain consistency across hundreds of pages. But the thoughts, logic, and voice are entirely my own. Each reflection began with lived experience — moments in real plants, real teams, and real failures that taught what alignment truly means.

The machine assisted in structure; the meaning came from the field. It allowed me to capture ideas at the speed they occurred, to test phrasing against the very principles this book teaches — precision, coherence, and intent. The goal was never to let technology think for me, but to let it think with me, mirroring the process of inquiry that defines evidence-based leadership itself.

Every page has been reviewed, edited, and approved by me to ensure it aligns with the purpose of this work: to merge human leadership with system discipline. The message is unchanged, whether typed by hand or accelerated by algorithm: Safety is not a script to memorize — it's a mindset to master.

This book stands as proof that technology, when guided by experience and anchored by purpose, can amplify clarity without diluting conviction. The thinking remains human. The evidence remains real. The alignment remains intentional.

— *Andy Page Ph.D.*

www.ingramcontent.com/pod-product-compliance
Lightning Source LLC
Chambersburg PA
CBHW062101270326
41931CB00013B/3174